Ascites

Colloquium Digital Library of Life Sciences

This e-book is a copyrighted work in the Colloquium Digital Library—an innovative collection of time saving references and tools for researchers and students who want to quickly get up to speed in a new area or fundamental biomedical/life sciences topic. Each PDF e-book in the collection is an in-depth overview of a fast-moving or fundamental area of research, authored by a prominent contributor to the field. We call these e-books Lectures because they are intended for a broad, diverse audience of life scientists, in the spirit of a plenary lecture delivered by a keynote speaker or visiting professor. Individual e-books are published as contributions to a particular thematic series, each covering a different subject area and managed by its own prestigious editor, who oversees topic and author selection as well as scientific review. Readers are invited to see highlights of fields other than their own, keep up with advances in various disciplines, and refresh their understanding of core concepts in cell & molecular biology.

For the full list of published and forthcoming Lectures, please visit the Colloquium homepage: www.morganclaypool.com/page/lifesci

Access to Colloquium Digital Library is available by institutional license. Please e-mail info@morganclaypool.com for more information.

Morgan & Claypool Life Sciences is a signatory to the STM Permission Guidelines. All figures used with permission.

Colloquium Series on Integrated Systems Physiology: From Molecule to Function to Disease

Physiology is a scientific discipline devoted to understanding the functions of the body. It addresses function at multiple levels, including molecular, cellular, organ, and system. An appreciation of the processes that occur at each level is necessary to understand function in health and the dysfunction associated with disease. Homeostasis and integration are fundamental principles of physiology that account for the relative constancy of organ processes and bodily function even in the face o f substantial environmental changes. This constancy results from integrative, cooperative interactions of chemical and electrical signaling processes within and between cells, organs and systems. This eBook series on the broad field of physiology covers the major organ systems from an integrative perspective that addresses the molecular and cellular processes that contribute to homeostasis. Material on pathophysiology is also included throughout the eBooks. The state-of the-art treatises were produced by leading experts in the field of physiology. Each eBook includes stand-alone information and is intended to be of value to students, scientists, and clinicians in the biomedical sciences. Since physiological concepts are an ever-changing work-in-progress, each contributor will have the opportunity to make periodic updates of the covered material.

Published titles

(for future titles please see the website, www.morganclaypool.com/page/lifesci)

Ascites
Jens H. Henriksen and Søren Møller
www.morganclaypool.com

ISBN: 9781615045662 paperback

ISBN: 9781615045679 ebook

DOI: 10.4199/C00085ED1V01Y201305ISP039

A Publication in the

COLLOQUIUM SERIES ON INTEGRATED SYSTEMS PHYSIOLOGY: FROM MOLECULES TO FUNCTION TO DISEASE

Lecture #39

Series Editor: D. Neil Granger, LSU Health Sciences Center, and Joey P. Granger, University of Mississippi Medical Center

Series ISSN
ISSN 2154-560X print
ISSN 2154-5626 electronic

Ascites

Jens H. Henriksen
University of Copenhagen, Denmark

Søren Møller
University of Copenhagen, Denmark

COLLOQUIUM SERIES ON INTEGRATED SYSTEMS PHYSIOLOGY: FROM MOLECULE TO FUNCTION TO DISEASE #39

MORGAN & CLAYPOOL LIFE SCIENCES

ABSTRACT

This volume deals with the history, aetiology, pathophysiology, symptoms, signs, prognosis, and rational treatment of ascites. During the past decade, our knowledge of the pathophysiology of ascites has increased substantially and more specific therapies are now based on aetiology and pathophysiology. It is the intention of this book to review recent progress in pathophysiology of ascites and therapies based on pathophysiology. Although the different types of ascites have a different aetiology and very different pathophysiology, the development of fluid in the peritoneal cavity is always a bad clinical sign. It has a severe prognosis, which is mainly dependent on the aetiology and progression of the underlying disease. However, among patients with ascites, the prognosis may be very different, mainly owing to the presence of portal venous hypertension, malignancy in the abdominal cavity, and end-stage congestive heart failure. The addition of complications like the hepatorenal syndrome and bacterial peritonitis, whether spontaneous or secondary, adds heavily to the bad prognosis. Since hepatic ascites are by far the most complex with respect to pathophysiology, complications, and treatment, emphasis is put on the description of this entity. Ascites of other aetiologies are mentioned along with hepatic ascites, in particular, if the pathophysiology differs from ascites of hepatic origin.

KEYWORDS

ascites, Budd-Chiari Syndrome, cardiac dysfunction, cirrhosis, cirrhotic cardiomyopathy, chronic liver disease, hepatic nephropathy, hepatorenal syndrome, hyperdynamic circulation, kidney injury, liver failure, malignant ascites, neuroendocrine dysfunction, portal hypertension

Contents

Abbreviations

ACE	Angiotensin converting enzyme
AQP2	Aquaporine-2 water channels
AVP	Arginine vasopressin
BV	Blood volume
CAMP	Cyclic adenosine monophosphate
CB	Canabinoid receptor
CBV	Central and arterial blood volume
$CC1_4$	Carbon tetrachloride
CGRP	Calcitonin gene-related peptide
CO	Cardiac output
COX	Cyclooxygenase
ET-1	Endothelin-1
HR	Heart rate
GFR	Glomerular filtration rate
HRS-1	Hepatorenal syndrome type-1
HRS-2	Hepatorenal syndrome type-2
HVPG	Hepatic venous pressure gradient
INR	International normalised ratio of clutting
LVEF	Left ventricular ejection fraction
MAP	Mean arterial pressure
NSAID	Non-steroidal anti-inflammatory drug
P_A	Ascitic fluid hydrostatic pressure
P_C	Capillary hydrostatic pressure
π_A	Ascitic fluid colloid osmotic pressure
π_p	Plasma colloid osmotic pressure
PRA	Plasma renin activity
PPCD	Post-paracentesis circulatory dysfunction
TIPS	Transjugular porto-systemic shunt
PV	Plasma volume
PRA	Plasma renin activity
RAAS	Renin angiotensin aldosterone system

RBF	Renal blood flow
SBP	Spontaneous bacterial peritonitis
SNS	Sympathetic nervous system
SVR	Systemic vascular resistance
V2	Vasopressin-2 receptors

CHAPTER 1

Introduction

Normal fluid homoeostasis includes dynamic shifts in water, crystalloids, and proteins between the various compartments of the body [1-3]. The fluid dynamic is controlled by refined mechanisms that include water and solute intake, renal handling of water and solutes, hemodynamic/colloid osmotic forces, and neurohumoral regulation [4, 5]. Fluid retention is a characteristic feature in the progressive state of chronic liver disease and other diseases, where the accumulation of fluid in the peritoneal cavity occurs in a substantial number of patients [5-8]. Over the last decade, much effort has been devoted to the role of vasodilation, abnormal blood volume distribution, hyperdynamic cardiovascular decompensation, and kidney dysfunction in the formation of especially hepatic ascites [9-15]. A considerable advance in the understanding of the microvascular dynamics and pathophysiology of the extravascular volume regulation has been achieved by studies with indicators of different molecular size [5, 16-18]. Advances in the insight of the dynamics of lymphatics and the peritoneal space have, in part, been obtained by studies on patients undergoing peritoneal dialysis [19, 20]. Renal dysfunction in patients with advanced cirrhosis has been intensively studied over several decades [7, 21-24]. The present book will deal with the regulation of the extracellular fluid volume, aetiology, ascites pathogenesis, and other aspects of the pathophysiology of the ascitic syndrome and complications to ascites. Some cornerstones and recent advances in the understanding of the normal state, especially with respect to the regulation of transvascular exchange at the local level, are also considered.

Ascites is a pathophysiological condition with increased fluid in the intraperitoneal space. Normally, a few millileters of fluid with a composition similar to that of the interstitial space is present between the visceral and parietal serous membrane of the peritoneal space [25]. Under pathological conditions this amount of fluid can increase to several liters, and in patients with tense ascites, 20-30 l are regularly seen in untreated cases. Ascitic fluid is, in contrast to the minimal normal intraperitoneal fluid, often of a different composition compared to interstitial space fluid.

CHAPTER 2

Pathophysiology of Ascites

2.1 HISTORY

Ascites (from Greek *ascos* – a bag) has been known since ancient times [26, 27]. Hippocrates stated that: "When the liver is full of fluid and this overflows into the peritoneal cavity so that the belly becomes full of water, death follows." Others correlated the occurrence of dropsy and ascites with shortness of breath, mild cough, and loss of appetite. On account of thirst, much cold water is swallowed and the fluid is transferred to the peritoneum. It was also said that: "Ascites begins from the liver, the water unnaturally collected here, unless evacuated, injures both the liver and all the rest of the internal organs." Abdominal paracenthesis was performed in ancient Rome and it was said: "Avoid by all means a sudden evacuation, for some ignorant persons having evacuated the vital spirit with the fluid, have immediately killed the patient." The morbid fluids were classified by the Greeks into *tympanities*, *leukophlegmasia* (or *hyposarca*), and *ascites,* as described by Jarcho (27). The latter was considered to be relieved more easily in slaves than in freemen, since hunger and thirst could be easily constrained in the former. The patient should walk much, run a little, and his upper part in particular was to be rubbed while he held his breath. Food should be of the mild class and a dry wine may be beneficial, if it is very thin. Moreover, it is good to measure, every day with a string, the circumference of the abdomen.

To summarize the knowledge of ancient Rome and Greece: Abdominal effusions may complicate long-standing diseases, especially heart failure, malaria, and excessive alcohol intake. The patient should drink only enough to sustain life. Diuretic drinks are the best; the quantity of ingested fluid and quantity of urine should be measured daily in addition to the circumference of the abdomen (26, 27). Abdominal paracentesis should not be sudden. Most of these observations and therapeutic initiatives are still useful in clinical routine today. An illustration of ascites and paracenthesis from Nicalas Tulp 1641 is shown in Figure 2.1.

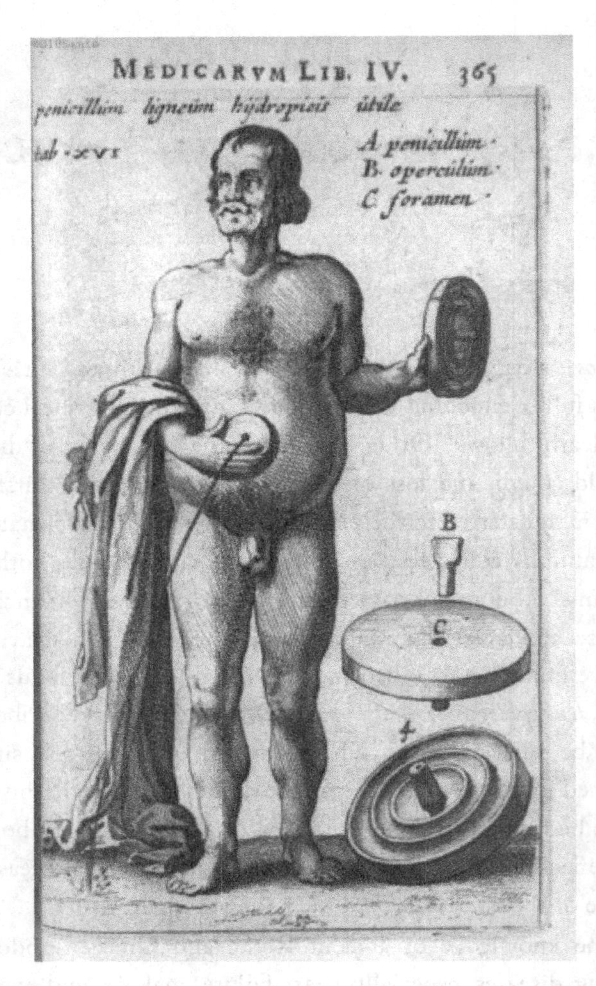

Figure 2.1: Illustration of ascites from *Observationes Medicinae*, a masterpiece by the Dutch Physician and Anatomist, Nicolaes Tulp, the professor known from the famous dissection lesson painted by Rembrandt. Tulp's *Observationes Medicinae* is from 1641, dedicated his son Peter Tulp, who tragically died young. In Book 4, Chapter XLII, 17 *penicillium ligneum* and *hydrops peritonei* illustrate fluid accumulation in the peritoneal space.

The anatomist Thomas Bartholin described the lymphatics in intestines and liver in an executed criminal who had had a meal enriched in lipids an hour before the execution. In the 16th century, the Oxford physiologist, Richard Lower, described experimental ascites in this way (Lower R: 1679): "The right side of the chest having been pierced between the seventh and eight ribs a little below the region of the heart, one should touch the site of the vena cava, and then push the side as close as one can to the vein so that one has greater ease in passing the thread around and tying it tightly.... Scarcely is the operation done than the dog begins to languish and dies a few hours after. When one does the dissection, one finds in the abdomen a considerable quantity of serosity, a type of dropsy known as ascites." With some refinements this type of experiment was repeated by Ernest Starling in the late 19th century and by modern researchers in the second part of the 20th century (28).

In 1892-1896, Ernest H. Starling, William M. Bayliss, J.B. Leathes, and the surgeon A.H. Tubby investigated the formation of lymph and absorption of fluid from the serous cavities [29-33]. The essential point is that Starling and Bayliss give qualified estimates of changes in hydrostatic pressure of the capillaries based on changes in arterial and venous pressures in different vascular beds. Obstruction of the aorta brought about a fall in arterial blood pressure, almost no change in venous blood pressure, and a fall in the capillary hydrostatic pressure with reduced formation of lymph as the outcome. In contrast, obstruction of the inferior vena cava above the liver veins causes a fall in arterial pressure, but a substantial rise in venous pressure, and a similar rise in the hydrostatic pressure in the liver capillaries, but only a small rise or no change in hydrostatic pressure in the intestinal capillaries, see Figure 2.2.

The result was a highly increased formation of lymph with a high content of protein. This is the first quantitative explanation of the difference between liver lymph and intestinal lymph, and of the importance of pressure and protein dynamics in ascites formation. In 1896, Starling introduced the significance of colloid osmotic pressure of albumin with the words: "The importance of these measurements lies in the fact that although the osmotic pressure of the proteins of the plasma is so insignificant, it is of an order of magnitude comparable to that of the capillary pressures; and whereas capillary pressure determines transudation, the osmotic pressure of the proteins of serum determines absorption, so that at any given time there must be a balance between the hydrostatic pressure of the blood in the capillaries and the osmotic attraction of the blood for the surrounding fluids. Here we have the balance of the forces necessary to explain the accurate and speedy regulation of the quantity of fluid." [33] This clear-sighted, highly imaginative, and very impressive interpretation of complicated and contradictory results was not given a favorable reception and certainly not accepted a hundred years ago. However, this turned out to be so, and it was later recognized by Eugene M. Landis and August Krogh in the late 1920s (28). Today, most researchers accept these important elements in the capillary factors of ascites formation. In addition to local factors, the systemic fluid balance with water-salt intake and renal sodium-water handling and neuro-hormonal regulatory mechanisms are now recognized.

Today, the term *ascites* means an abnormal and widespread (general) accumulation of fluid between the two layers of the peritoneum. The definition therefore excludes more localized small fluid accumulations as found in local inflammatory processes, for example seen in patients with appendicitis and other local peritoneal inflammation.

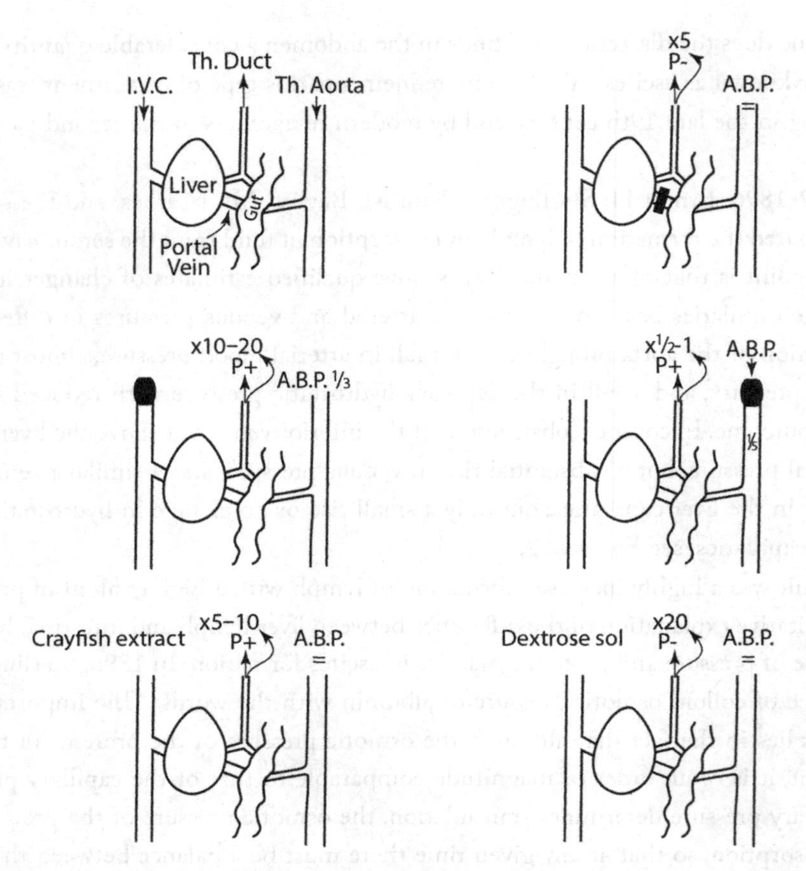

Figure 2.2: Illustrations of Starling's experiments. The effect of occlusion of the portal vein, occlusion of the inferior vena cava (I.V.C.), occlusion of the thoracic aorta, a first-class lymphagogue (toxic crayfish extract) and a second-class lymphagogue (Dextrose solution) upon lymph flow from the thoracic duct, duct lymph protein content, and arterial blood pressure (A.B.P.). The figures above the thoracic duct are the multiplier of the initial lymph flow; P+ and P- denote duct lymph richer or poorer in protein, respectively. The figure below A.B.P. is the multiplier of the initial arterial blood pressure, = denotes little change from the initial arterial blood pressure. From Henriksen [28].

2.2 AETIOLOGY OF ASCITES

The most common diseases that cause ascites are: liver cirrhosis, heart failure, abdominal malignancy, peritoneal infection, hypoalbuminemia, and pancreatitis; see Table 2.1. Besides ascites of the well-known aetiologies, clinicians may meet patients with a more obscure genesis, like nephrogenic ascites, ascites in patients with acute viral hepatitis, alcoholic hepatitis, Meig's syndrome, B-avitaminosis, tropical diseases, and chylous disorders.

Table 2.1: Ascites: aetiology and diagnosis of common clinical ascites

Classification of Ascites	Aetiology	Verification of Diagnosis and Aetiology
Hepatic	Cirrhosis/steatosis Budd-Chiari syndrome Veno-occlusive disease	Ultrasonography, hepatic vein catheterization, paracentesis; demonstration of inferior vena cava-hepatic vein thrombosis; absence of vein thrombosis, liver biopsy, pressure measurement
Malignant	Abdominal primary and secondary malignancy	Ultrasonography, CT-scan, MR-scan, cytology, laparoscopy
Cardiac	Right-side heart failure (chronic constrictive pericarditis)	Demonstration of heart disease (ultrasonography of inferior vena cava and heart, CT-scan, liver biopsy: passive congestion), hepatic vein/right heart catheterisation
Infectious	Peritonitis, vira, bacteria, fungi, spontaneous subacute peritonitis in liver disease	Ascites bacteriology (by tuberculosis: ultrasonography and laparoscopic biopsy)
Hypoalbuminaemic	Nephrotic syndrome Malnutrition (kwashiorkor, beri-beri)	Serum albumin < 20–25 g/l (300–350 μmol/l) Urinary albumin excretion
Pancreatogenic	Acute pancreatitis, chronic pancreatitis, cysts	Ascites amylase concentration; faecal elastase-1 test; ultrasonography, CT-scan

Nephrogenic ascites may develop during chronic renal failure in patients undergoing haemodialysis. Some cases are part of a generalized fluid retention or concomitant congestive heart failure, but in other cases no obvious cause can be identified [25]. There may be a relatively high ascitic protein concentration, and altered sodium transport has been suggested. Meig's syndrome is the combination of a benign ovarian tumor, ascites, and hydrothorax. It has been reported that local factors like lymphatic obstruction in the tumor may induce ascites production [34]. In malnutrition and B-avitaminosis, right-side congestive heart failure is often present. Tropical diseases like filariasis may cause ascites by lymphatic obstruction. However, it should be noted that the common presinusoidal portal hypertension in schistosomiasis does not give rise to ascites, except in special cases where hypoalbuminemia is also present. Congenital abnormalities of abdominal lymphatics, lymphomas, disseminated malignancy, and traumatic or surgical damage of lymphatics may result in chylous ascites, although rarely (less than 0.5%).

2.3 DIAGNOSIS OF ASCITES

History, clinical examination, measurement of body weight, ultrasonography, CT-scan, and diagnostic puncture of the abdominal wall are the essentials in the clinical diagnosis of ascites. A patient with tense ascites is shown in Figure 2.3. Less serious, but often overlooked, other causes of abdominal distension should be remembered: pregnancy, distended urinary bladder, giant cysts (ovarian, congenital), pancreatic pseudocysts, large retroperitoneal masses, intraperitoneal fat, and *pseudomyxoma peritonei* are not infrequently mistaken as ascites or vice versa. Abdominal ultrasonography and diagnostic paracenthesis will most often clear out the situation.

Physical examination is not sufficiently sensitive to detect minor amounts of intraperitoneal fluid. Verification by paracenthesis cannot be recommended for that purpose alone, as a negative result may represent a failure of the puncture, and a positive fluid aspiration may be derived from puncture of for example an ovarian cyst. At least 500 ml of ascitic fluid must be present before fluid can be withdrawn during paracenthesis [25]. Abdominal ultrasonography may detect as little as 50 ml fluid if the patient is investigated in different positions. All patients with a clinical suspicion of ascites should therefore have performed a complete abdominal ultrasonography. Clues to the aetiology of ascites may often be found by this examination. CT-scans may also be useful. The liver, intrahepatic vessels, spleen size, and portal vein (patency, flow direction, size) may give useful information as to aetiology.

Although the diagnosis of ascites and its aetiology may often be determined by fairly simple means, mistakes are not uncommon. In this context a complete catheterization of the large veins, hepatic veins, and right heart with pressure measurements is a useful tool; see Figures 2.4 and 2.5a-g. Elevated pressure in the right site of the heart indicates cardiogenic ascites, and a significant pressure gradient between the right atrium and the inferior vena cava, characteristic of intradiaphragmatic causes, is absent. Hepatic ascites is followed by an elevated hepatic sinusoidal pressure gradient (wedged-to-free hepatic vein pressure or portal pressure more than 5 mmHg above the pressure of the inferior vena cava). Differential diagnostics related to the Budd-Chiari syndrome (hepatic vein obstruction), can usually be solved by this catheterisation procedure, eventually together with an inferior cavography. In this context, it should be mentioned that the ascitic fluid hydrostatic pressure is often similar to that of the inferior vena cava, but of cause always a little below the latter value (1-2 mmHg). Only in cases with obliteration of the upper part of the inferior vena cava due to, for example, thrombosis or tumor infiltration, may the hydrostatic pressure substantially exceed that of the ascitic fluid, thereby giving differential diagnostic information.

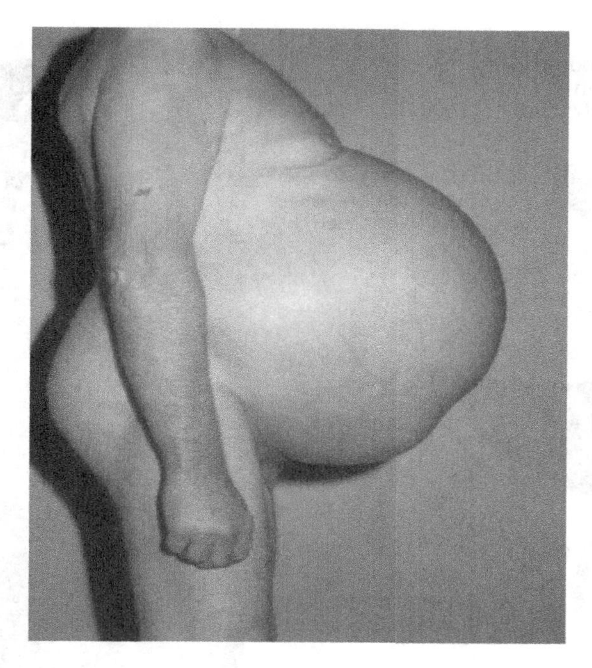

Figure 2.3: A male patient with tense ascites. Courtesy by professor Flemming Bendtsen, MD, Gastro Unit, Medical Devision, Hvidovre University Hospital.

In selected cases diagnostic laparoscopy may be performed. This procedure was originally introduced to diagnose peritoneal carcinosis and tuberculosis. Laparoscopy may verify the presence of peritoneal fluid, disclose its aetiology, and allow elective biopsies and therapy like removal of benign tumors, ovarian cyst and, closure of any leaks.

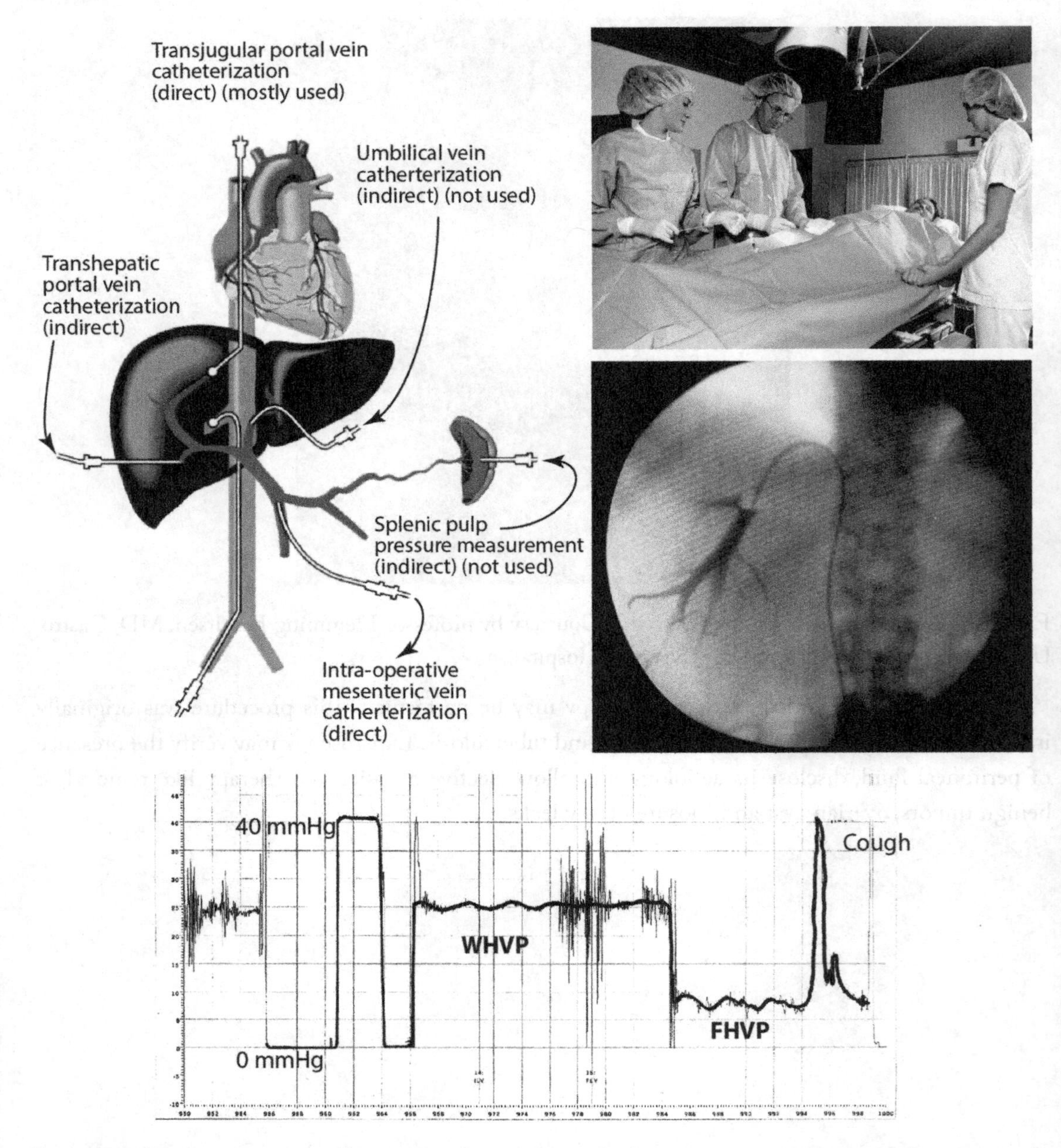

Figure 2.4: During an invasive liver vein catheterization, the hepatic venous pressure gradient is measured as the difference between the wedged hepatic venous pressure(WHVP) and the free hepatic venous pressure (FHVP) under flouroscopic control. In this patient the hepatic venous pressure gradient was approximately 25 – 8 = 17 mmHg.

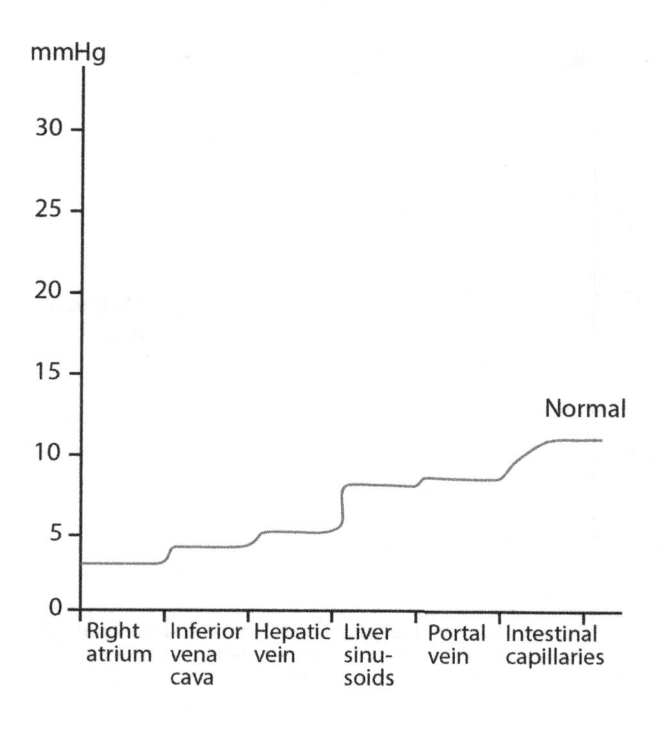
mmHg
30
25
20
15
10
5
0
Normal
Right atrium
Inferior vena cava
Hepatic vein
Liver sinusoids
Portal vein
Intestinal capillaries

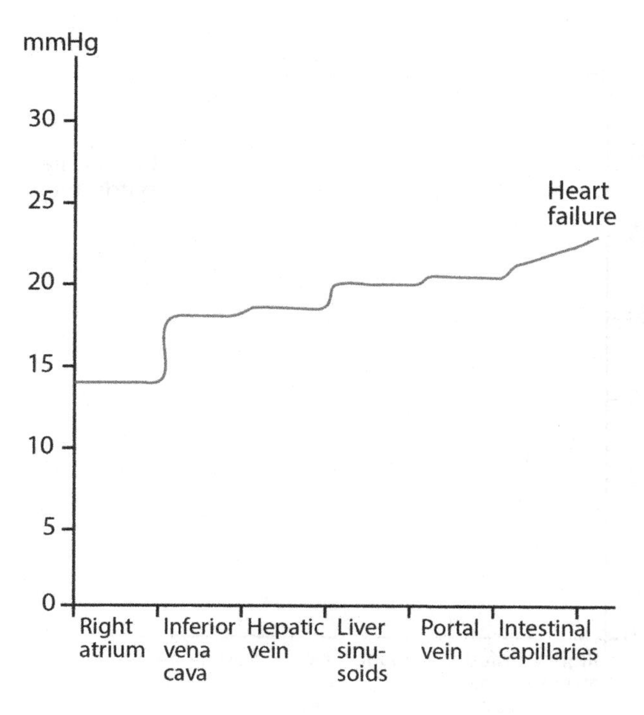
mmHg
30
25
20
15
10
5
0
Heart failure
Right atrium
Inferior vena cava
Hepatic vein
Liver sinusoids
Portal vein
Intestinal capillaries

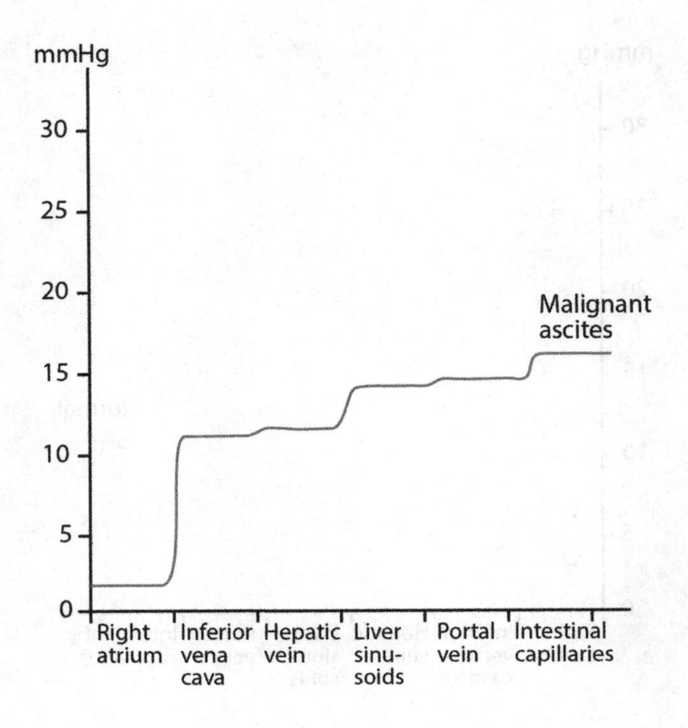
mmHg
30
25
20
15
10
5
0
Malignant ascites
Right atrium
Inferior vena cava
Hepatic vein
Liver sinusoids
Portal vein
Intestinal capillaries

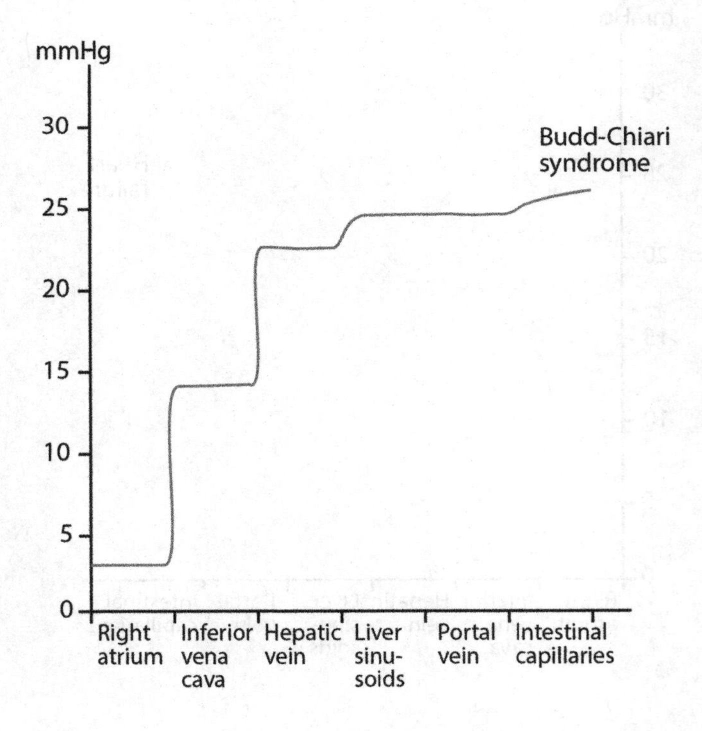
mmHg
30
25
20
15
10
5
0
Budd-Chiari syndrome
Right atrium
Inferior vena cava
Hepatic vein
Liver sinusoids
Portal vein
Intestinal capillaries

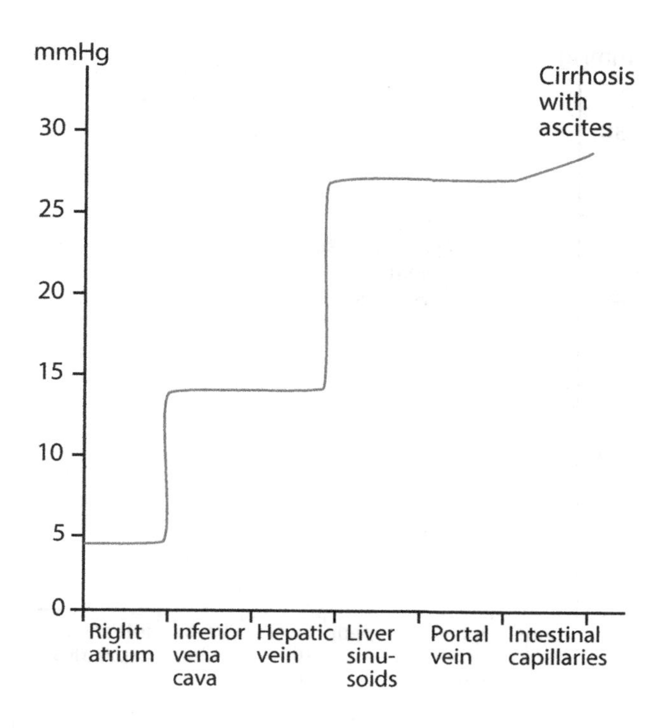
mmHg
30
25
20
15
10
5
0
Cirrhosis with ascites
Right atrium
Inferior vena cava
Hepatic vein
Liver sinu-soids
Portal vein
Intestinal capillaries

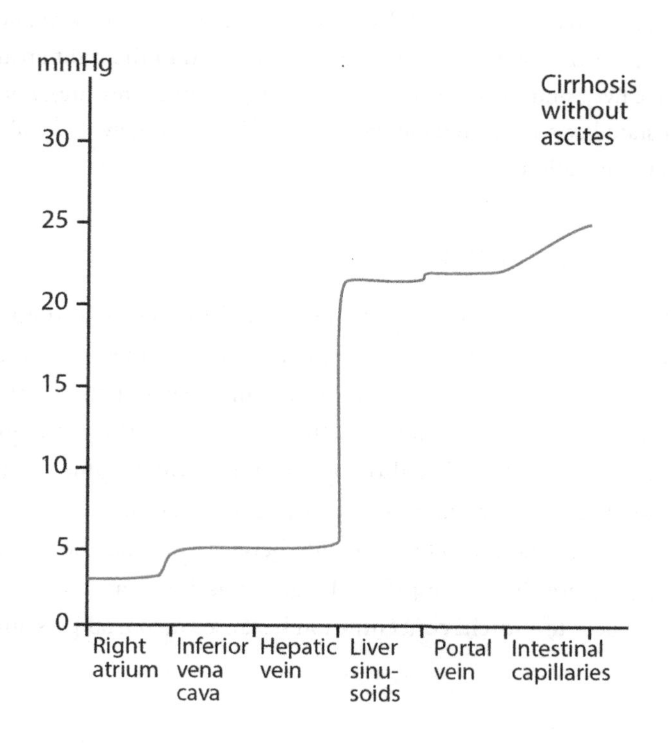
mmHg
30
25
20
15
10
5
0
Cirrhosis without ascites
Right atrium
Inferior vena cava
Hepatic vein
Liver sinu-soids
Portal vein
Intestinal capillaries

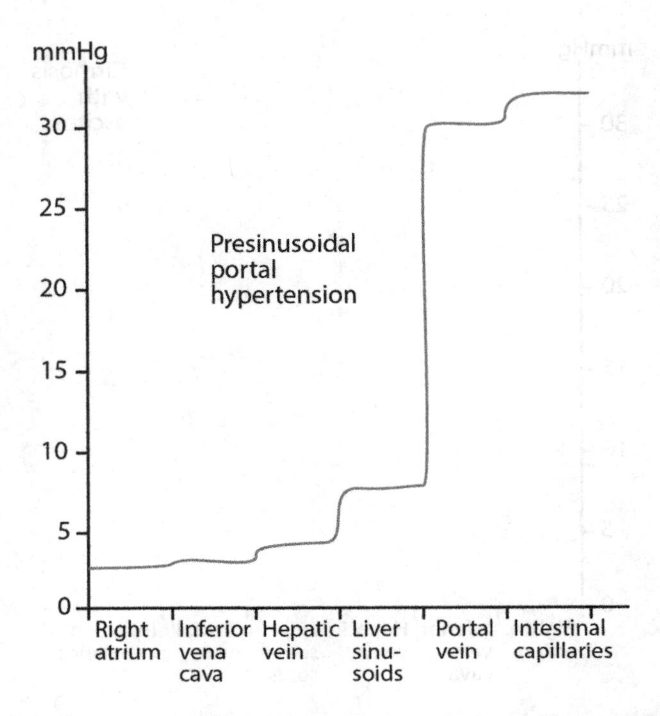

Figure 2.5a–g: Splanchnic pressure profiles in the normal condition and in several pathological states, including heart failure, malignant ascites, Budd-Chiari syndrome, cirrhosis with and without ascites, and presinusoidal portal hypertension. The pressures are illustrated in the right atrium, the inferior vena cava, hepatic veins, liver sinusoids (equal to wedged hepatic vein pressure), portal vein. and intestinal capillaries. Each state has a characteristic pressure profile, which may be applied in the diagnosis during hepatic vein catheterisation.

2.3.1 ANALYSIS OF ASCITIC FLUID

Diagnostic information by paracentesis may be important. This is due to the increasing incidence of spontaneous bacterial peritonitis (SBP). Determination of protein and albumin concentration is only meaningful if these concentrations are determined in plasma at the same time [35, 36]. The concentrations are influenced by several factors such as the level of the portal pressure, actual state of fluid balance, and treatment given. This also applies to electrolytes, glucose, and specific density. For pathophysiological studies it is of interest that the plasma-ascites concentration, difference of albumin, or the difference in colloid osmotic pressure between plasma and ascitic fluid may mirror the portal hydrostatic pressure difference [36]. Malignant ascites may therefore be different from cirrhotic ascites with respect to protein concentration because the portal pressure gradient is low in the former mentioned condition, where a higher protein concentration is present in the ascitic fluid under the assumption that plasma protein is normal.

Leucocyte count with fraction of polymorph nuclear leucocytes and bacterial culture are important diagnostic tools, which should always be performed in all patients with ascites. In spontaneous subacute bacterial peritonitis, the leucocyte count is above 250,000 per ml, and bacteriological verification and antibiotic sensitivity should always be determined before treatment is initiated. Cytological examination for malignant cells may be relevant when the patient is suspected of malignant ascites, hepathoma, and peritoneal carcinosis. However, it should be remembered that the absence of malignant cells does not exclude abdominal malignancy.

Measurement of triglyceride and amylase concentrations is indicated in chylous ascites and acute pancreatitis. Spontaneous presence of blood in ascitic fluid usually indicates malignancy, but may also be traumatic, tuberculous, or caused by haemorrhagic diathesis and clotting disorders.

2.3.2 MEASUREMENT OF ASCITIC FLUID VOLUME

The amount of accumulated intraperitoneal fluid may be between 50 ml and 20-30 l, and in a few cases even more than 30 l. The exact determination of the size of the ascitic fluid volume has limited value in daily clinical practice, where measurement of body weight will give a clue as to ongoing fluid accumulation or fluid loss. However, in research and clinical cases where measurement may be of interest, several methods are at hand: evacuation of fluid by abdominal paracentesis, DEXA-scan of the abdomen with determination of lean, CT-scan with specific algorithms, ultrasonography with specific algorithms, and measurement with the indicator dilution technique.

Evacuation by paracentesis usually leaves a substantial amount of fluid in the interperitoneal space (at least 0.5 l). However, as total paracentesis has become a valuable therapeutic procedure this may be the most often applied method for semiquantitative or near quantitative determination of ascitic fluid volume.

Estimation by compiling slices from CT-scans (or ultrasonography or MRI) gives reproducible values (coefficient of variation down to 4%), but these procedures are not applied very often [37].

DEXA-scans with regional body composition co-determines the water volume in the abdominal wall and intestines [38]. This procedure is fast, gives a minimum radiation, and has no patient discomfort. The precision and accuracy is better than that obtained by evacuation by paracentesis, but it demands a high-resolution DEXA-scanning equipment of good quality, calibrated for medium/large body size.

The most accurate and classic method is volume determination by quantitative injection of an indicator into the ascitic fluid and subsequent sampling here [39]. The indicator should be a protein bound tracer, for example iodine-labeled human serum albumin or a protein bound dye-like indocyanine green. After the quantitative injection the abdomen is carefully kneaded and samples of ascitic fluid are taken 20-60 min after the indicator injection. The reproducibility is good, and the amount of indicator absorbed by the abdominal lymphatics during the first hour is small and can be ignored. The precision of this method is about 3% (coefficient of variation) [39, 40].

2.4 PHYSIOLOGY OF LIVER, GASTROINTESTINAL TRACT, AND PERITONEAL SPACE VASCULATURE

Filtration of fluid from microcirculatory vessels is a normal physiological process that aids the exchange of substances between plasma and tissue [41, 42]. Normal fluid homeostasis is controlled by refined mechanisms that include haemodynamic and colloid osmotic forces, dynamic shifts of proteins between the different compartments in the body, and neurohumoral regulation. The balance between intravascular and extravascular fluid is governed by the Starling forces, i.e., capillary and interstitinal fluid hydrostatic pressure, plasma and interstitial fluid colloid osmotic pressure, microvascular permeability, hydraulic conductivity, and lymphatic drainage [42, 43].

Endothelium of continuous capillaries have small pores (paracellular clefts) of 0.8-2.0 nm in diameter that allow passage of water and small solutes. In addition, there are specific water transporters of the aquaporin system. Larger pores, 20-30 nm in diameter, allow leakage of proteins with molecular diameters approaching 10-15 nm. The gastrointestinal tract has fenestrated capillaries with basement membrane. Sinusoids with aggregations of fenestrae, measuring 100 nm in diameter, and no basement membrane are found in the liver and spleen [5]; see Figure 2.6. The microcirculation in the liver consists of two vascular beds: The hepatic sinusoids and the peribiliary capillary plexus. The latter gets blood from the hepatic artery, whereas the sinusoids receive their blood supply from a mixture of portal venous and hepatic arterial circulation, as well as from the outflow of the peribiliary capillary plexus (25).

Surplus capillary filtrate from the liver and gastrointestinal tract drains via capsular and hilar spaces through hepatic and intestinal lymphatics into the thoracic duct [44, 45]. The peritoneal space is chiefly drained through lymphatics on the abdominal side of the diaphragm to the right lymphatic duct [5, 42]. If there is an imbalance between the filtrative forces and those which drain the peritoneal space, surplus fluid will appear and sequestration of fluid into the peritoneal cavity leads to the formation of ascites.

In the heart, kidney, and liver disease, fluid retention with the formation of edema is a characteristic feature in the progressive stage, and in cirrhosis, the accumulation of fluid in the peritoneal cavity occurs in a substantial number of patients with advanced disease. Before a detailed outline of the pathophysiology of ascites is given, the normal vasculature in the liver, intestine, and peritoneal wall will be described.

Normal:
Fenestrated capillary
without basement membrane

Cirrhosis:
Defenestration without
reduction in pore size

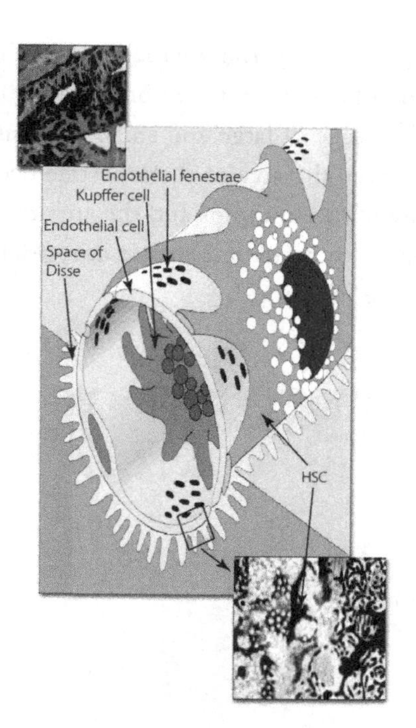

Figure 2.6: Schematic illustration of a liver sinusoid. The endothelial cells are of the fenestrated type without basement membrane. The fenestrae are scattered in sieve plate formations, which cover an area of approximately 15-20% in the normal condition. Perivascular cells surround the endothelial cells, but the sinusoid is in close contact with the hepatocyte with its numerous micro-villi. In cirrhosis there is a defenestration of the liver sinusoids with occurrence of basement membrane (so-called capillarization), and the number of fenestra are substantially reduced without significant reduction in size of the single fenestra (around 100 nm in diameter). The net result is a reduction of transvascular hydraulic conduction, without significant sieving.

2.4.1 NORMAL VASCULATURE

The normal hepatosplanchnic vascular bed is a low-pressure system with a small drop in pressure from the intestinal capillaries to the portal vein. A portal venous pressure of about 8-10 mmHg and a small hepatic venous pressure gradient (HVPG) of 2-5 mmHg across the sinusoids of the liver is considered normal [5].

From a structural point of view, the normal blood lymph-barrier consists of the capillary endothelial cell, a basement membrane to which the endothelial cells are attached, occasionally pericytes, which wrap around the capillary (in the gastrointestinal tract and peritoneal membrane, but not in hepatic or splenic capillaries), connective tissue spaces, and it ends with the terminal lymphatics [35, 42, 46]. The microcirculation in the normal liver has discontinuous capillaries (sinusoids) without a basement membrane [46-50]. The fenestrae in the sinusoidal lining allow almost unrestricted passage of large and small proteins in the same proportion (bulk flow) [5, 35]. In the normal condition, the area of the fenestrae occupies about 10-15% of the capillary wall [35]. The porous nature of the sinusoids and the architecture of the liver give an impression of a microvasculature designed for rapid and almost unrestricted transport of large amounts of fluid, solutes, and metabolites; see Figure 2.7.

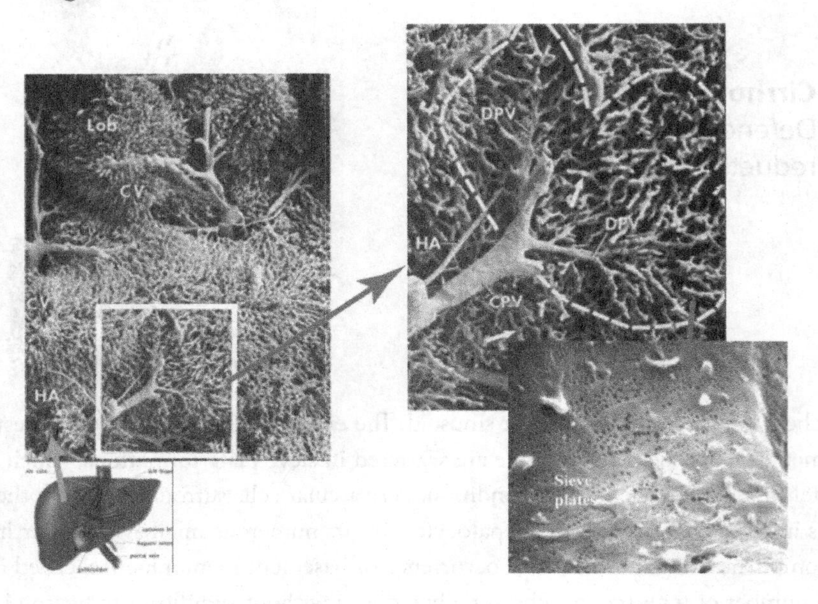

Figure 2.7: The anatomic location of the microvasculature in the liver has been studied with different cast in animals and man. In the presence of cirrhosis, a microvascular reduction in crosssectional area will result in an increased hepatic vascular resistance that is mainly located at the central vein or post-sinusoidally from a functional point of view.

The exchange of material is either diffusive (i.e., governed by differences in concentration) or filtrative/convective (i.e., governed by sieving and fluid volume flow, owing to differences in pressure) [5]. Small molecules (sodium, chloride, glucose, lactate, etc.) and gases (oxygen, carbon dioxide) are transported through the entire capillary wall almost solely by simple diffusion or facilitated by carriers [5]. In the liver, this transport rate is large and almost only limited by the hepatic plasma flow rate [5, 51]. The transport of large molecules (proteins) is in the main filtrative/convective, but, at very low rates of lymph flow, diffusion may be perceptible. Permselectivity, i.e., preferential transport of smaller molecules owing to restriction of larger molecules in the fenestrae, is only present for very large proteins [5, 52]. Studies by Goresky and co-workers with albumin [mol. wt 69,000,

Stokes-Einstein (S-E) diameter 7 nm] have shown that the sinusoidal-persinusoidal communication is normally free, i.e., limited only by plasma flow [48, 52, 53]. This is well understood from our present knowledge of the morphology of the liver sinusoidal lining. Electron microscopic studies confirm that restriction (permselectivity) can be present only for very large molecules with an S-E diameter above approximately 20 nm [35, 50, 52]. Thus, in the normal liver there is close agreement between results obtained by morphometric and by physiological methods. The perisinusoidal space (space of Disse) may therefore be regarded as a paravascular part of the plasma volume. This concept is further stressed by the fact that the conventionally determined plasma volume (indicator dilution technique) includes this paravascular space in the liver [48, 51, 53, 54].

The transport of fluid and solutes from the perisinusoidal space into portal connective tissue and subcapsular spaces is not clearly defined from a morphological point of view [35, 44, 55]. However, most physiological studies with molecules of different sizes indicate no appreciable permselectivity here, except at very low rates of lymph flow [5, 53]. The transport from portal and subcapsular interstitium to terminal lymphatics is considered to take place by bulk carriage [35, 44]. Owing to the large size of the fenestrae in the liver sinusoids and the bulk flow of fluid with a high lymphatic content of protein, no colloid osmotic pressure gradient exists across the sinusoids or the liver blood-lymph barrier [44], and it is possible that liver tissue is the barrier, limiting the movement of plasma proteins from blood to lymph.

An exception is the peribiliary capillary plexus, which has continuous type capillaries with a much lower permeability than the normal liver sinusoids [53]. Filtration of fluid with a low protein content from the peribiliary capillary plexus is the reason why the normal liver lymph-plasma protein ratio is not 1.00, but rather 0.95 [16, 44]. The significance of the peribiliary capillary plexus in normal fluid dynamics is probably small and will not be considered further in this review.

The microcirculation in the gastrointestinal tract has a fenestrated capillary wall with a basement membrane that is relatively impermeable to macromolecules, but very permeable to water and smaller molecules [47, 52]. Therefore, only part of the plasma colloid osmotic pressure is exerted across the microvascular endothelium. Published values of the transmicrovascular colloid osmotic pressure gradient for the stomach, small intestine, and colon range from 11.5–13.0 mmHg [5, 48]. Intestinal capillaries show permselectivity with characteristic sieving [16]. The intestinal lymph-plasma protein ratio is normally relatively low [5, 16] in accordance with the relatively tight capillary membrane here. Normally, fluid from the gastrointestinal interstitium is drained to intestinal lymphatics and further into the thoracic duct, and fluid, solutes, and proteins are only to a very limited extent transported into the peritoneal space.

The blood capillaries in the peritoneal membrane are the continuous type with a basement membrane. Over the last decade a number of studies on transperitoneal dynamics have been performed in uremic patients undergoing peritoneal dialysis [20, 56]. The essentials of these studies are that both diffusive and filtrative transport participates in the overall transperitoneal dynamics. The peritoneal clearance of low molecular substances, such as glucose and creatinine, is in the order of 5-15 ml/min. Whereas fluid and low molecular solutes exhibit a major direct transperitoneal transport, proteins and other high molecular substances are in the main transported back into the bloodstream via the subdiaphragmatic lymphatics [5, 57-59].

The Starling equation describes the transcapillary dynamics [28, 60]:

$$J_v = K_f [P_{mv} - P_t) - \sigma(\pi_p - \pi_t)],$$

where J_v is the filtration rate, K_f the filtration coefficient, P_{mv} the microvascular hydrostatic pressure, P_t the tissue hydrostatic pressure, σ the osmotic reflection coefficient, π_p the microvascular (plasma) colloid osmotic pressure, and π_t the tissue colloid osmotic pressure. The filtration coefficient (K_f) is a measure of the hydraulic conductance of the microvascular barrier between plasma and tissue. K_f is determined by two major factors: water permeability and the surface area available for exchange. The reflection coefficient (σ) describes the fraction of the effective colloid osmotic pressure generated across the microvasculature. Impermeable proteins have a σ value of 1, since they generate 100% of their osmotic pressure difference. Conversely, small freely permeable proteins ($\sigma = 0$) do not generate any osmotic pressure difference. Values of σ indicate that 92%, 85%, and 78% of the total plasma colloid osmotic pressure gradient is generated across the microcirculatory barrier in the small intestine, colon, and stomach, respectively, but that 0% is generated in the liver [5, 44, 61, 62]. In microcirculatory beds in which σ for protein is > 0, the colloid osmotic pressure gradient across the microvascular barrier is an important determinant of fluid dynamics. For example, a fall in πp, caused by a reduction in the concentration of plasma proteins, can disturb the Starling forces and lead to an accumulation of interstitial fluid in amounts too large to be cleared by the lymphatics.

2.5 HEPATIC ASCITES

2.5.1 MICROVASCULATURE IN SINUSOIDS AND CAPILLARIES

Sinusoidal hypertension is an essential feature in the hepatic ascites syndrome. Various theories have been put forward as to its genesis [63-66]. A decrease in the overall hepatic vascular cross-sectional area, an increase in vascular resistance located postsinusoidally or close to the outlet of the sinusoids, and an increase in mesenteric inflow are important elements [64]. The hydrostatic pressure in the sinusoids is increased in cirrhosis and in Budd-Chiari syndrome. Because the hepatic and splanchnic microcirculations are in series, events occurring in the liver are transmitted upstream to the splanchnic organs in the form of increased portal and intestinal capillary pressure. A small presinusoidal pressure component has been described in some patients with alcoholic liver disease, but most investigators agree on the concept, that sinusoidal and portal pressure are increased to the same degree in cirrhosis [8, 45, 67]. Thus, the elements in the Starling equation favor an outward movement of fluid in portal-sinusoidal hypertension [5, 44].

The sinusoidal lining is altered in cirrhosis [35, 68, 69]. Most studies indicate a tightening of the sinusoidal wall. During the progression of cirrhosis, the sinusoids of the liver become transformed into less permeable capillaries, and some sinusoids develop a continuous, defenestrated endothelium with development of a basement membrane and even pericytes. This transformation of the capillary barrier directly affects the rate of which fluid and protein filter from the intravascular plasma into the surrounding tissue, as described by the Starling equation. Schaffner & Popper first described the so-called capillarization with appearance of a basement membrane and collagenization of the perisinusoidal space in 1963 [68, 70], and later studies on morphology have confirmed the observations [5, 71]. These results, combined with newer investigations on protein kinetics, indicate a reduction in the number of fenestrae rather than in their size [18, 35, 72]. This implies that the sinusoidal permselectivity remains small in cirrhosis, and consequently the effective colloid osmotic pressure gradient across the sinusoids is also small [45]. However, the tightening of the sinusoidal wall and collagenization of the perisinusoidal space decrease the hydraulic conductivity, but these alterations in the liver blood-lymph barrier only represent a limited counterweight against the increased sinusoidal pressure.

Tracer kinetic studies have shown a substantially increased blood-to-lymph transport of water, solutes, and proteins with increased trans-sinusoidal filtration and highly elevated hepatic lymph flow [5, 18]. Direct measurements on surgically exposed lymph vessels in cirrhosis have thus shown an enlarged thoracic duct with a lymph flow 5–15 times higher than normal values (2–4 liters) [45, 73]. When the lymph drainage keeps pace with the enhanced filtration, the spillover of fluid into the peritoneal cavity is minimal and the patient will be free of ascites [35, 45], but when the trans-sinusoidal filtration exceeds the transport capacity of the lymph vessels, surplus fluid will pass into the peritoneal space. The rate of transport into the peritoneal space is, however, relatively small (< 5%) compared with that of lymph directly entering the thoracic duct [5, 35, 57].

Tracer kinetic studies with protein indicators of different molecular sizes have indicated that most of the intraperitoneal protein comes from the liver via trans-sinusoidal filtration [5, 20, 35, 44]. Concentrations of small and large proteins in proportion to their level in plasma as well as equal relative transport rates suggest bulk carriage, i.e., flow through openings without restriction [5, 35, 44, 72[. The liver (and spleen) is the only location that can provide this type of transport. In contrast, earlier studies by Witte et al. concluded that in advanced cirrhosis ascitic fluid is predominantly generated by the intestines [73]. This may be true for the water and non-protein components of the ascitic fluid, but protein kinetic studies have shown that trans-sinusoidal filtration is the main origin of ascitic fluid protein [5, 72, 74].

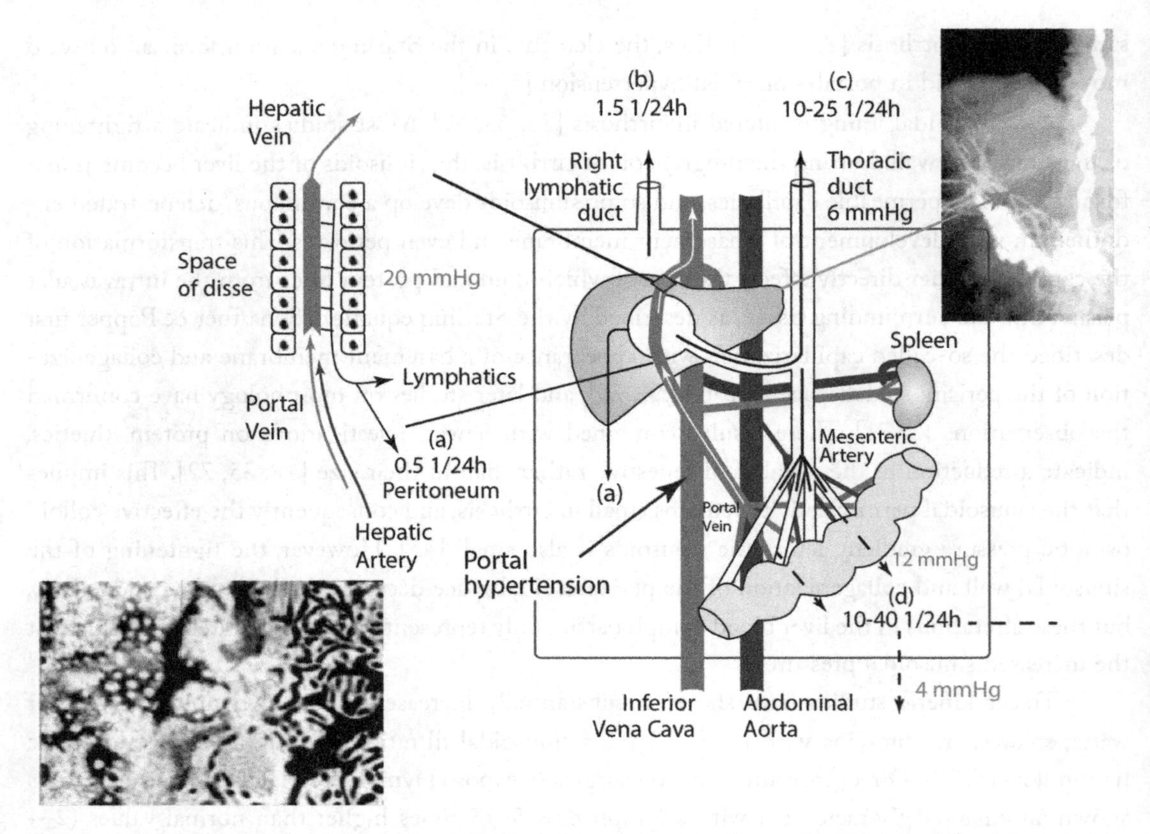

Figure 2.8: A survey of fluid dynamics in liver and gastrointestinal tract in patients with cirrhosis with sinusoidal-portal hypertension. Enhanced trans-sinusoidal filtration of fluid, solutes and protein take place with subsequent drainage mainly into hilar and subcapsular lymphatics. The flow in the thoracic duct is highly increased as compared to 2-4 l/24 h in normal subjects. As long as drainage keeps pace with the increased formation of filtrate, the peritoneal space is free of ascitic fluid. With imbalance in filtration and re-absorption a small amount is spilled over into the peritoneal cavity from the liver. This fluid with high protein content will equilibrate, according to the Starling forces with the large capillary area in the gastrointestinal tract, and ascitic fluid is formed with lower protein content, depending on the size of the portal pressure and the hydrostatic pressure in capillaries in the gastrointestinal tract. Direct transfer of fluid and low molecular solid takes place over the peritoneal lining with an overall transport rate equivalent to a clearance value around 10-30 ml per minute. The transport of low molecular solids is in the main diffusive. Transport of proteins and other high molecular solutes are in the main filtrative and returns to plasma by lymphatic drainage.

Figure 2.8 illustrates the different transport kinetics from plasma to ascitic fluid and back, and Table 2.2 summarizes clearance rates from plasma to ascitic fluid of substances with different molecular weights. Figure 2.9 shows transport curves from different tracers.

Table 2.2: Plasma-to-peritoneal clearance of substances with different molecular weight in patients with hepatic ascites and patients with uraemia

	Peritoneal clearance (ml/min)	
	Uraemia	Hepatic ascites
Sodium ($^{24}Na^{+}$, 24+)	16#	44 ± 4.0
Ethanol (46+)		43#
99mTc-MDP (271+)	9.1 ± 4.0	
51Cr-EDTA (324+)	13.1 ± 3.8	26 ± 3.5
125I-Albumin (69,000+)	0.11 ± 0.03	0.23 ± 0.076

Mean ± SD
+ Molecular weight in Dalton.
The clearance values are measured according to the kinetic theory as transport rate into the peritoneal cavity devided by average plasma concentration of indicator.
Measurements from a few patients.
Most values are higher in patients with hepatic ascites compared to patients with uraemia.
From Henriksen et al. [5, 18], Joffe & Henriksen [19, 79].

In patients with cirrhosis, increased hydrostatic pressure in the hepatic and splanchnic microcirculations constitutes a major element in the formation of ascites. Increased intravascular hydrostatic pressure also increases the filtering surface area of the microcirculation through two major mechanisms: First, increased pressure promotes the opening of non-flowing capillaries, which can then contribute to fluid filtration. Second, a mechanism that might not be readily apparent from Starling's equation is an increase in the length of filtering vs. absorbing sections of the capillary bed [5, 60].

In cirrhosis, there is a significant vasodilation of lymphatics in the intestine and an increase in the number of mesenteric lymphatic vessels. The lymph-plasma ratio of proteins in the intestine is reduced from 0.60 to 0.18. Thus, colloid osmotic pressure in the interstitium is decreased to an even greater extent than in plasma. Considering that permeability to plasma proteins in the intestine stays relatively constant during cirrhosis, it is evident that the lymphatic dilution of protein is the result of greatly enhanced capillary fluid filtration.

The gastrointestinal transvascular filtration in portal hypertension is thus dominated by a protein-poor filtrate, which will equilibrate with the protein-rich fluid coming from trans-sinusoidal filtration [18]. On the assumption that lymph provides an accurate reflection of the contents of interstitial fluid, the colloid osmotic pressure gradient across the microvascular barrier can be estimated from lymph and plasma either by using an osmometer or by measuring the protein concentration and subsequently applying equations that relate protein concentration to colloid osmotic pressure. Basal filtration coefficients may vary by a factor of 20 among the splanchnic organs and liver [5, 64, 75]. These variations may represent true differences in microvascular permeability or the exchange surface area, or may contain some technical artefacts of the evaluations of K_f. σ is approximately 0.8–0.9, whether the portal hypertension is acute or chronic [5]. Intestinal capillary permeability does not appear to be substantially increased by either acute portal hypertension or cirrhosis [76, 77].

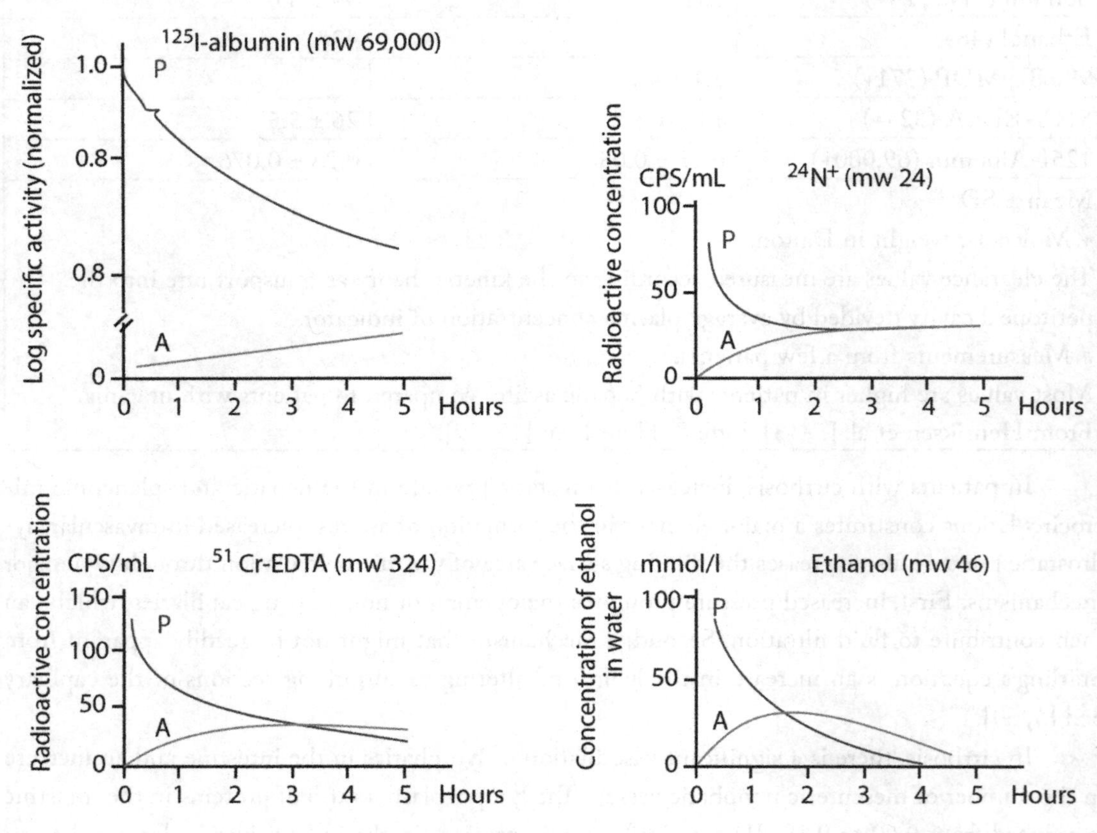

Figure 2.9: Radioactive and chemical concentrations in plasma (P) and ascitic fluid (A) after intravenous injection of substances with different molecular weights. Transport rates can be determined from these curves. A net transport from plasma into ascitic fluid takes place within the first hours after intravenous injection. It is seen that high molecular substances are transported much slower than low molecular substances. The high molecular proteins are transported by a filtrative mechanism, whereas low molecular substances, like sodium, ethanol, and Cr-EDTA, are transported by both filtration and diffusion.

2.5.2 TRANSPORT FROM THE PERITONEAL CAVITY TO THE BLOODSTREAM

The dynamics of transperitoneal transport has recently been studied extensively in uremic patients on peritoneal dialysis [19, 20, 56, 74]. In many ways, the transport kinetics in these patients is similar to that of patients with cirrhosis. In cirrhosis, ascitic fluid is always equilibrated with the surrounding interstitial tissue and plasma, which gives it the same osmolality as that of the plasma [75]. This is because of the effect of the small solute crystalloid osmotic pressure. As the Donnan effect across the peritoneal membrane is small, no substantial electric potential difference exists across this membrane. It should, however, be kept in mind that different substances are transported from ascitic fluid back into plasma by different routes and by different mechanisms [35].

2.5.3 TRANSPORT OF HIGHLY SOLUBLE SUBSTANCES WITH LOW MOLECULAR WEIGHT

An example is ethanol (mol. wt 46), which crosses all biological membranes easily [74]. The transport rate is fast (transperitoneal clearance is about 20–50 ml/min), and a relatively slow equilibration is seen only in patients with a very large ascitic volume [5, 74]. Transport takes place through the entire peritoneal membrane. Diffusion is by far the dominating transport process (see Table 2.2).

2.5.4 TRANSPORT OF LOW MOLECULAR EXTRACELLULAR SUBSTANCES

These are plasma electrolytes, creatinine, glucose, ^{51}Cr-EDTA, etc. These substances have a transperitoneal clearance of 5–45 ml/min [74, 78, 79]. They are transported by diffusion as well as by filtration [74]. The main transport takes place across the entire peritoneal surface area. As the concentration of a number of these endogenous substances is relative stable over time, a significant net transport takes place only during accumulation and removal of ascitic fluid (net reabsorption, peritoneovernus shunting, or paracentesis).

2.5.5 TRANSPORT OF HIGH MOLECULAR SUBSTANCES

As mentioned above, the ascitic origin of most of these substances (albumin, transferrin, immunoglobulins, lipoproteins, etc.) is the result of trans-sinusoidal filtration. The magnitude and mechanisms of protein transport from the ascitic fluid back into the plasma have been studied by introducing different protein tracers into the ascitic fluid, followed by blood and ascitic fluid sampling [5, 20, 35, 44, 57, 59]. Occurrence of a protein tracer in plasma after a substantial time lag, an almost linear rise in plasma concentration (Figure 2.10), and a similar rate of transport of smaller and larger protein indicators suggest bulk flow transport through tubes [19, 75]. This is consistent with animal experiments, which have shown that the peritoneal cavity is mainly drained by lymphatics on the abdominal side of the diaphragm [56, 59]. The transport rate in the right lymphatic duct is relatively small [5]. It should, however, be stressed that some protein kinetic studies, especially those performed in uremic patients and in animals, have indicated that there is also a significant protein transport directly into the interstitial space of the peritoneal membrane [20]. This agrees with recent kinetic results showing that some of the protein loss in patients on chronic ambulatory peritoneal dialysis may come from the interstitial space next to the peritoneal membrane [19]. Whether this holds true in patients with cirrhosis is not known at present. In cirrhosis most results indicate that the main transport of ascitic fluid protein takes place through the lymphatic route [19, 56]. However, the overall transport rate of ascitic fluid protein back into the plasma is relatively low (clearance about 0.05-0.2 ml/min; see Figures 2.8 and 2.10). The capacity of this transport of ascitic fluid protein back into plasma is the rate-limiting step for steady ascitic fluid reabsorption (see below).

The general porosity of the capillary membrane is currently under debate. Two-pore and three-pore models have been proposed [20, 41]. The latter includes facilitated transport of water by specific transport proteins (aquaporin) and possibly also an albumin transporter (albondin) [80, 81]. However, the importance of theses specific transport mechanisms has not been established in the peritoneal cavity, but their quantitative role is most likely to be minor [20, 80].

2.5.6 DYNAMICS OF LOCAL ELEMENTS IN THE FORMATION AND THERAPY OF HEPATIC ASCITES

The presence of protein in the peritoneal space is an important element for fluid sequestration here. Reabsorption of isotonic protein-free fluid is complete and relatively quick, owing to the effect of the plasma colloid osmotic pressure [5]. When protein is present in the peritoneal space (as in cirrhosis), fluid and solute movements take place in the different areas, as described above, and ascitic fluid is formed as a result of the Starling forces. The protein concentration in cirrhotic ascitic fluid is much lower than that in plasma because of the hydrostatic/colloid osmotic equilibration over the large surface area of the gastrointestinal tract [5, 36, 74, 78, 82]. Accordingly, the colloid osmotic pressure in ascitic fluid is close to that of the intestinal lymph [73]. As intestinal blood capillaries are relatively impermeable to protein, increased gastrointestinal capillary hydrostatic pressure produces a filtrate low in protein concentration [36].

Thus, ascitic fluid may be considered to be a mixture of protein-rich fluid from trans-sinusoidal filtration in the liver and protein-poor fluid from transcapillary gastrointestinal filtration, the mixing ratio being governed by hydrostatic/colloid osmotic forces [5]. Consequently, the colloid osmotic pressure of the ascitic fluid will decrease, but the effective colloid osmotic pressure gradient (i.e., plasma minus ascitic fluid colloid osmotic pressure) will increase and thereby counteract the filtration force of the elevated portal-intestinal capillary pressure (the so-called protein "wash-down" effect) [5, 78]. The effective colloid osmotic pressure gradient thus plays a role in the dynamics of transperitoneal/transintestinal fluid, but the size of the colloid osmotic pressure gradient is governed by the magnitude of the transmural hydrostatic pressure, i.e., the portal pressure [35, 83, 84]. Hence, the effective colloid osmotic pressure gradient may be looked upon as a mirror image of the portal venous pressure; see Figure 2.11.

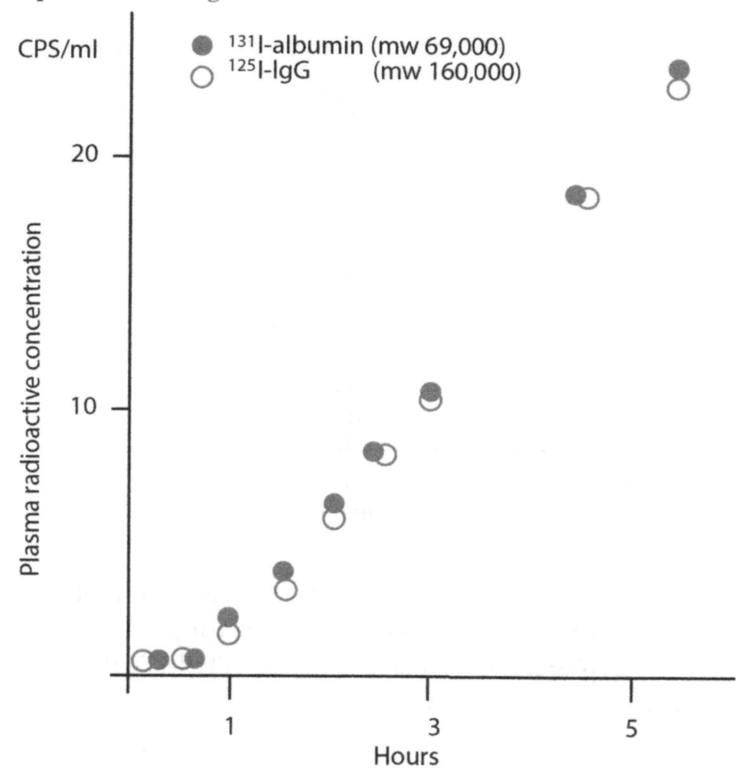

Figure 2.10: Time activity curves in plasma after intraperitoneal injection of different radio-labeled protein indicators. The plasma radioactivity curves show a characteristic delay of 5-45 min before they rise in an almost straight line with the same velocity of small and large molecular proteins, indicating transport through tubes, i.e., lymphatics.

Owing to the low capacity of ascitic fluid-to-plasma transport of proteins, this transport becomes the rate-limiting step in the amelioration of ascites [83]. During intensive diuretic treatment plasma colloid osmotic pressure may increase and portal pressure may be reduced (plasma volume contraction), which results in a decreased rate of ascites formation and increased transperitoneal transport of water and crystalloids back into plasma [20, 85]. However, the effective colloid osmotic pressure gradient cannot become higher than the portal pressure, which stresses the presence of protein in the peritoneal cavity as a crucial element in fluid accumulation [78].

Transport rates to and from the peritoneal space are not only of academic interest, they are also important for therapy. Thus, for the reasons mentioned above, mobilization of fluid by diuretic treatment should be adjusted to the rate of inflow into the peritoneal cavity/rate of protein transport back into plasma [75, 78].

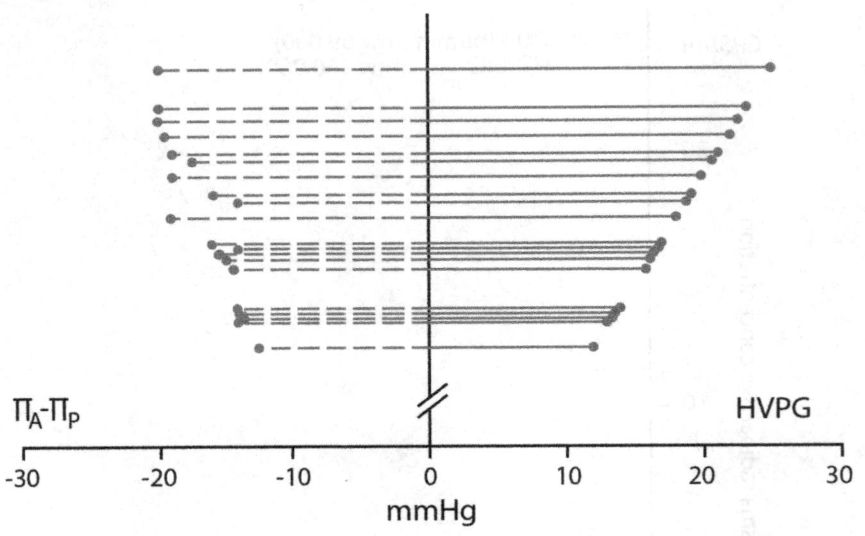

Figure 2.11: Effective colloid osmotic pressure (ascitic fluid minus plasma colloid osmotic pressure) illustrated as a mirror image of the transmural portal pressure in patients with cirrhosis. π_A: ascitic fluid colloid osmotic pressure, π_P: plasma colloid osmotic pressure, WHV-IVCP: wedged hepatic venous minus inferior vena cava pressure (= portal pressure).

The slow peritoneal lymph drainage can be accelerated by surgical implantation of an artificial "megalymphatic," i.e., peritoneovenous shunt [86, 87]. The astonishing ameliorating effect of this therapy on ascites illustrates the importance of efficient drainage from the peritoneal cavity. However, numerous complications and side effects render the therapy of limited clinical value [86, 87].

Another important way of removing ascitic fluid and peritoneal protein is by paracentesis [88]. A number of controlled clinical studies have proved the efficacy of this treatment (for overview, see [88]). Owing to the transport dynamics described above, paracentesis particularly repeated paracentesis, should never be performed without giving the patients a plasma expander, as this procedure will otherwise lead to a reduction in the effective plasma volume and vasodilatation with postparacentesis circulatory dysfunction (PPCD) with potential major adverse effects [10, 11, 89, 90].

2.5.7 EFFECT OF VASODILATORS/CONSTRICTORS ON MICROVASCULAR FLUID DYNAMIC IN HEPATIC ASCITES

Acute portal hypertension alone is not enough to cause severe oedema or ascites, owing to "protein wash-down" and lack of peritoneal protein. However, splanchnic arteriolar vasodilation may play a crucial role in the formation of ascites, transmitting arterial pressure directly through the arterioles with decreased resistance to the filtering capillaries [91-94], which can deliver fluid, in combination with the liver supply of protein by trans-sinusoidal filtration. The autonomic nervous system and modulators with paracrine and autocrine activity (e.g., serotonin, nitric oxide, endothelins, substance P, vasoactive intestinal polypeptide, calcitonin gene-related peptide) may influence the microvascular fluid exchange [9, 91-93, 95-100]. This may be brought about by several mechanisms: recruitment of capillaries, altered balance in between pre- and postcapillary vascular tonus, change in microvascular permeability, and altered receptor/transporter status. Unlike several microvascular beds, the sinusoids of the liver probably do not exhibit recruitment, either in the normal or in the pathological state [45, 84]. Available evidence indicates that all sinusoids are open, although there may be a change in the contribution of portal venous/hepatic artery flow and some outflow regulation in different parts of the liver [49].

The sympathetic and parasympathetic nervous systems modulate the vascular tone in the splanchnic area and in the liver. This holds true in the normal state and in patients with cirrhosis and heart failure [4, 101-103]. There are indications that increased sympathetic nervous activity may raise portal venous pressure and α- and β-adrenoceptor blockers reduce portal and sinusoidal pressure [104-108]. In this way, autonomic tone may contribute to trans-sinusoidal and transcapillary gastrointestinal filtration. The role played by neurogenic vasodilators, such as the calcitonin gene-related peptide and substance P for transvascular dynamics, is not known.

Contractile elements in the lining of the sinusoidal fenestrae have been described in animal experiments and *in vitro* growth of sinusoidal endothelial and stellate cells has shown a modulating effect of different vasoactive substances [109-112]. Therefore, neurohumoral modulators may, at least in part, participate in the control of local trans-sinusoidal exchange. Capillary permeability increases in the presence of bradykinin and glucagons. Consequently, the leakage of plasma proteins caused by bradykinin and glucagons may reverse the lymphatic dilution of protein observed under the increased capillary filtration. The role of these mechanisms in pathophysiology is at present merely speculative, but it opens up the possibility of pharmacological intervention.

The universal vasodilator, nitric oxide, and the endothelial vasoconstrictors (endothelins) play an important role in the systemic and splanchnic circulatory changes in cirrhosis [92, 111, 113-116]. Their role is the local transvascular fluid dynamics in cirrhosis is, however, at present unclear.

2.6 PERITONEAL CARCINOSIS

Malignant ascites is found in patients with tumors in the female genitals (ovaries), and in cases with widespread tumor cells to the peritoneal membrane from other viscera. These metastases are followed by local neo-vascularization with formation of leaky capillaries [117, 118], which in turn gives a protein and fluid escape directly into the peritoneal cavity [119]. This leads to ascitic fluid with a high protein concentration. Plugging of the diaphragmatic lymphatic capillaries by malignant cells has been described in animal experiments as well as in human experiements [120]. In accordance with this, tracer kinetic studies indicate a reduced lymph drainage from the peritoneal cavity in carcinosis [5, 120]. The increased pressure found in the inferior vena cava and liver veins in this condition is merely secondary to accumulation of fluid in the peritoneal cavity (see Figure 2.5). Although comprehensive studies of the pathophysiological mechanisms behind malignant ascites are few, it seems reasonable to conclude that the combination of increased microvascular leakiness and lymphatic insufficiency are the major causative factors in this condition.

Tumors in the ovaries may secrete mucin or mucin-like material into the peritoneal cavity which may be osmotically active and give rise to fluid accumulation without involvement of increased microvascular permeability. In this case plugging of peritoneal lymphatics may also contribute to fluid accumulation in the peritoneal cavity.

2.7 PERITONITIS WITH ASCITES

Peritoneal fluid accumulation follows different types of peritoneal inflammation (see also Section 5.2). The pathogenesis of fluid accumulation in patients with peritonitis is probably similar to that of inflammatory oedema, i.e., capillary damage with increased plasma leakage [121, 122]. Especially in tuberculosis, large amounts of ascitic fluid may persist for a long time, and these patients may be misinterpreted as having other causes of fluid accumulation, especially malignant tumors.

2.8 PANCREATIC ASCITES

The ascites of chronic pancreatitis is caused by rupture of a pseudocyst (approximately two thirds of the cases) or a pancreatic duct (approximately one third) [123].

2.9 CHYLOUS ASCITES

This condition is caused by obstruction or rupture of intestinal lymphatics, by which route ingested lipids normally are transported to the blood stream. Chylous ascites may also be present in cirrhosis in a few cases. The diagnosis is confirmed by analysis of ascitic fluid and scintigrafphy with radioactive labelled colloids.

2.10 NEPHROGENIC ASCITES

Nephrogenic ascites is not an entity from an aetiological point of view, but may have elements of right heart failure, hypoproteinemia, fluid overload, peritoneal inflammation, and even lymphatic obstruction [123, 124].

2.11 CARDIAC ASCITES

Ascites due to cardiac decompensation is seen in patients with severe right-side insufficiency and associated with massive oedema [125]. The mechanism is most likely similar to that of cirrhosis in the sense that the liver sinusoidal pressure, as well as lymph flow and protein capillary leakage, are increased in patients with congestive cardiac failure [126-128]. However, signs indicating a changed microvascular permeability in this condition are absent [5, 13, 126], and the portal sinusoidal pressure is in general lower in cardiac failure than in cirrhosis (see Figure 2.5). As the venous pressure is increased to approximately the same level in the liver and at the inlet of the large lymphatic ducts of the thoracic veins, it is conceivable (but not demonstrated) that the rate of lymph drainage may be slower than in cirrhosis. It is generally agreed that implantation of a peritoneo-venous shunt (i.e., addition of an artificial megalymphatic) is of no value in patients with congestive heart failure, and this procedure should not be performed due to the risk of pulmonary oedema and aggravation of the heart disease [45, 86]. Ascites formation secondary to a primary renal sodium-water retention may be present in heart failure [129].

2.12 HYPOPROTEINAEMIC ASCITES

When plasma albumin concentration is low and plasma colloid osmotic pressure declines below approximately 15-20 mmHg oedema occurs [5]. The presence of clinically significant augmentation of interstitial fluid is preceded by an increasing effective colloid osmotic pressure gradient (i.e., the colloid osmotic pressure in the interstitial space decreases faster than that of the plasma). This is due to protein "wash-out" and "wash-down" mechanisms [5]. These mechanisms are known, together with increased interstitial fluid pressure and lymph formation and flow, as "oedema-preventing forces." However, when these counter-regulating mechanisms reach a certain limit (either due to decreased plasma colloid osmotic pressure or insufficient capacity of lymph drainage) they are insufficient to prevent accumulation of interstitial fluid, and oedema results. In general it is argued that the "oedema-preventing forces" are stronger in the splanchnic area compared to e.g., the lower limb [130, 131]. Nevertheless, if the plasma protein concentration becomes sufficiently low (albumin concentration below approximately 300–350 mol/l (20–25 g/l), ascites will appear even in patients with normal vascular pressures and capillary permeability [74, 131]. Thus, the pathogenic mechanism behind this type of ascites is similar to that of pitting oedema seen in hypoproteinemia.

CHAPTER 3

Systemic Elements in the Ascites Syndrome

3.1 ABNORMAL DISTRIBUTION AND REGULATION OF PLASMA VOLUME IN HEPATIC ASCITES

3.1.1 SPLANCHNIC AND PERIPHERAL VASODILATATION

Splanchnic vasodilatation in cirrhosis is well documented and generates increased transvascular filtration, owing to higher intracapillary hydrostatic pressure and, to some extent, also increased capillary permeability, because of the larger microvascular surface area [12, 73, 93, 132]. In contrast to the splanchnic circulation, the peripheral circulation may be variably vasodilated and vasoconstricted in patients with cirrhosis [114, 133-136]. The renal circulation is typically vasoconstricted, owing to a highly elevated sympathetic nervous tone and elevated angiotensin II and endothelin 1 [137-141]. In addition, intrarenal vasodilators may be disturbed [142]. In skeletal muscle, subcutaneous and cutaneous tissue there are indications of different vascular tone with areas with microvascular dilation and constriction [14]. The role of this heterogeneous pattern in the overall control of extravascular fluid volume is not clear at present, but elevated venous pressure and reduced plasma colloid osmotic pressure are most likely important in the declive oedema in patients with cirrhosis [10, 143]. The transition of fluid from plasma into the interstitial spaces depends highly on the overall fluid status and the distribution of the volume of the circulation medium [2, 5, 43, 144]. Albumin and other plasma proteins leave the intravascular space via a rather slow transcapillary movement [28]; see Figure 3.1 [145]. This means that colloid osmotic material, primarily introduced into the vascular space, is transferred slowly into the interstitial space with the end result that a little more amount of albumin is located in the interstitial space than in the vascular volume in the normal condition [5]. In patients with cirrhosis and cardiac ascites this distribution is reverse, owing to interstitial space protein wash-down caused by the increased transvascular and lymphatic fluid exchange [18].

The volume distribution is not even within the vascular space as arteries and capillaries contain much less of the circulating medium than do veins and the splanchnic vessels [5, 146]. In patients with cirrhosis, especially in the more advanced stages, there is an increase in total vascular compliance, and the arterial compliance is also increased in these patients [9, 11, 134, 147, 148]. This means that a volume load has a relatively small effect on the intravascular transmural pressure, and that the distribution of the added volume may be different in the various vascular beds (144).

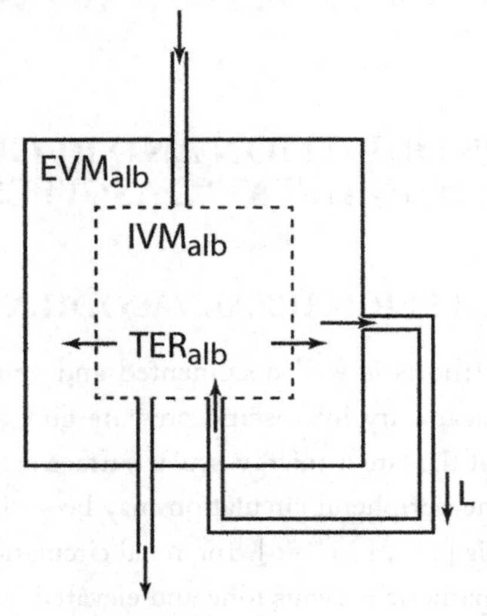

Figure 3.1a: Schematic illustration of transvascular albumin transport from plasma to interstitial space, and return back into the intravascular space by lymphatics. In addition synthesis and disposal are illustrated as arrows. IVM_{alb}: intravascular mass of albumin, TER_{alb}: transvacular escape rate of albumin, EVM_{alb}: extravascular mass of albumin.

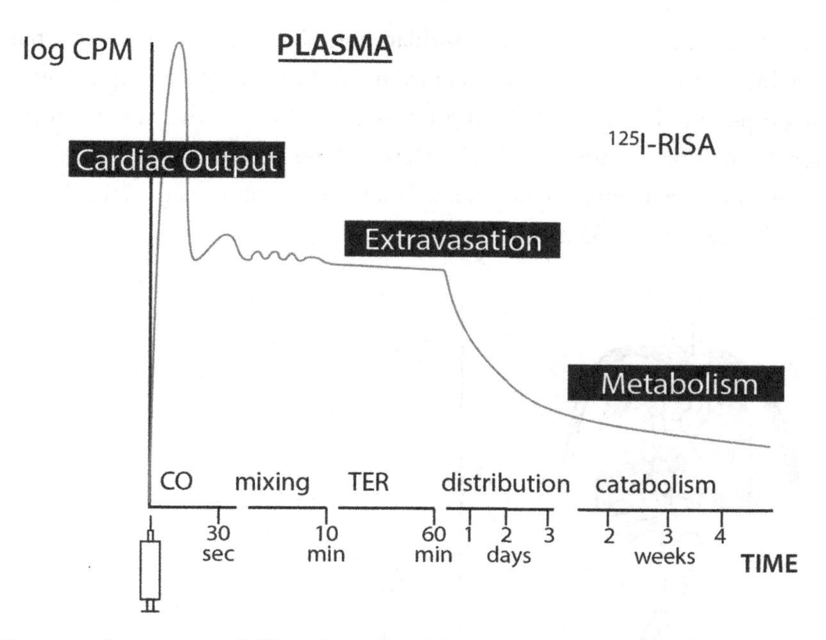

Figure 3.1b: Plasma radioactivity of ^{125}I-radioiodine-labeled serum albumine (125I-RISA) at different times after injection. This illustrates the influence of cardiac output (CO), intravascular mixing, transvascular escape rate (TER), distribution between extravascular and intravascular space, and irreversible catabolism, which takes place during the subsequent weeks.

Animal studies of experimental cirrhosis and portal hypertension, as well as investigations in patients with cirrhosis, have revealed that the blood volume in the splanchnic organs is substantially increased [21, 146]. Likewise, the overall blood volume is increased, but the central blood volume is decreased or normal in cirrhosis [149-151]. As shown by a dynamic indicator dilution technique, the central and arterial blood volume (that is the volume in the heart cavities, lungs, and central arterial tree) is decreased in patients with cirrhosis, especially in those with advanced disease [144]; see Figure 3.2 [151]. In agreement with the directly measured, reduced effective arterial blood volume, highly elevated homoeostatic markers, such as plasma renin activity (PRA), circulating catecholamines, aldosterone, and vasopressin, indicate that patients with advanced cirrhosis have functional hypovolaemia [3, 4, 138, 152, 153]. As plasma atrial natriuretic peptide may be normal or increased [154, 155], the overall picture points towards a situation with a reduced effective arterial blood volume and varying degrees of reduced central blood volume, depending on the absence or presence of cirrhotic cardiomyopathy [156-159]. The cause of the reduced effective arterial blood volume is largely the splanchnic vasodilation with reduced overall systemic vascular resistance and increased vascular compliance [3]; see Figure 3.3 [160, 161]. These findings are explained by the *arterial vasodilation theory*, which implies a reduced effective arterial blood volume and increased non-central blood volume consequent on vasodilatation [6, 13, 151]. Intimately related with the vasodilatation, which occurs before sodium retention and plasma volume expansion [91], is the increased cardiac output and low arterial blood pressure [14, 21, 147]. The cardiac output is increased, but not enough to maintain a normally high arterial blood pressure [144, 162]. The mechanisms counteracting the effective arterial hypovolaemia include, as mentioned, activation of all available vasopressor systems; see Figure 3.4a-b.

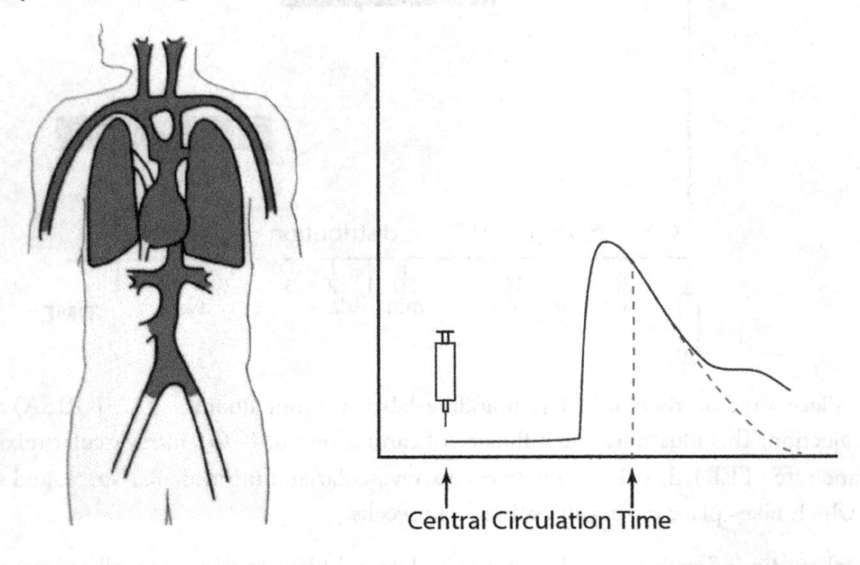

Figure 3.2a: The central and arterial blood volume (central effective blood volume) is the part of the circulation where volume and baroreceptors are located (i.e., the blood volume in the heart cavities, pulmonary circulation, and central arterial tree). Central and arterial blood volume (CBV) may be determined by injection of blood indicators and sampling at the aortic bifurcation. In this way, segmental blood volumes can be determined according to the kinetic theory as flow multiplied by circulation time.

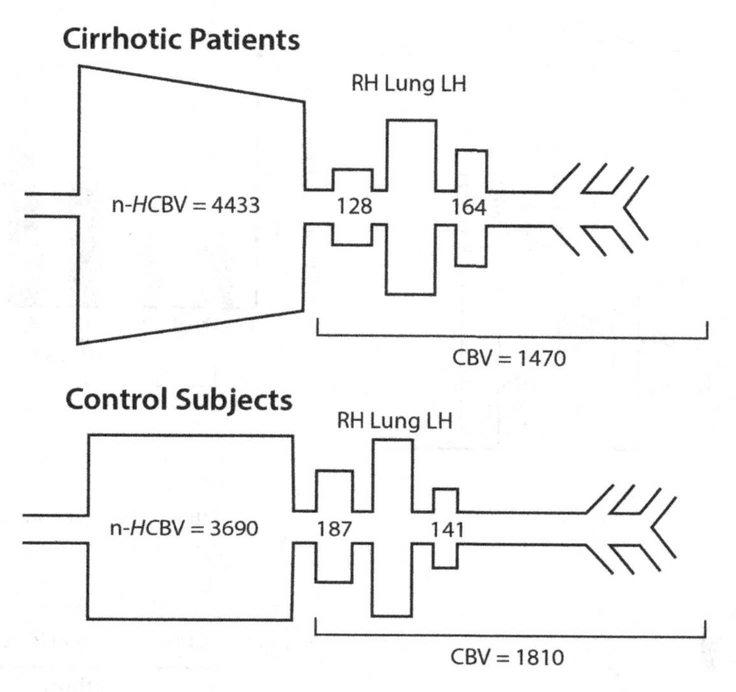

Figure 3.2b: Illustration of central and arterial blood volume (CBV) and non-central blood volume (n-CBV) in patients with cirrhosis and control subjects. It is seen that CBV is substantially reduced in patients with cirrhosis. However, CBV may contain sub-compartments, like the left ventricle (LH) and left atrium, where increased values may be found, depending on the presence of cardiac dysfunction (cirrhotic cardiomyopathy). The non-central blood volume and total blood volume/plasma volume are most often substantially increased in patients with cirrhosis. RH: right heart. From Møller et al. [150].

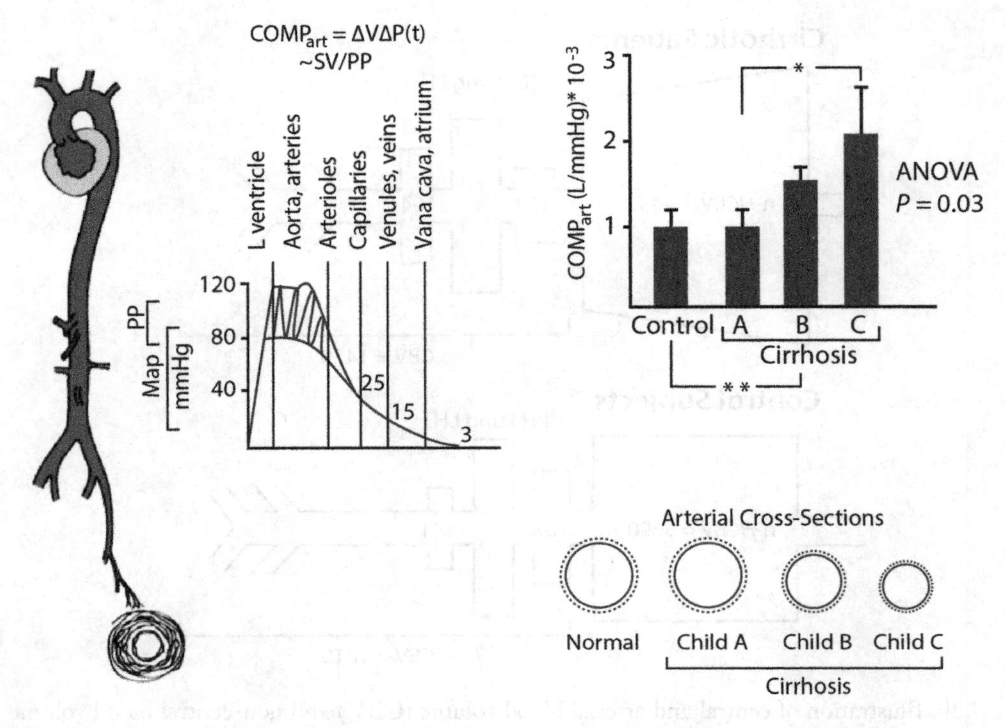

Figure 3.3: Arterial compliance (COMPart), i.e., change in intravascular volume (ΔV) relative to change in transvascular pressure (ΔP(t)) may be estimated as stroke volume (SV) relative to pulse pressure (PP). In patients with cirrhosis there is increasing arterial compliance in patients with advanced disease and ascites.

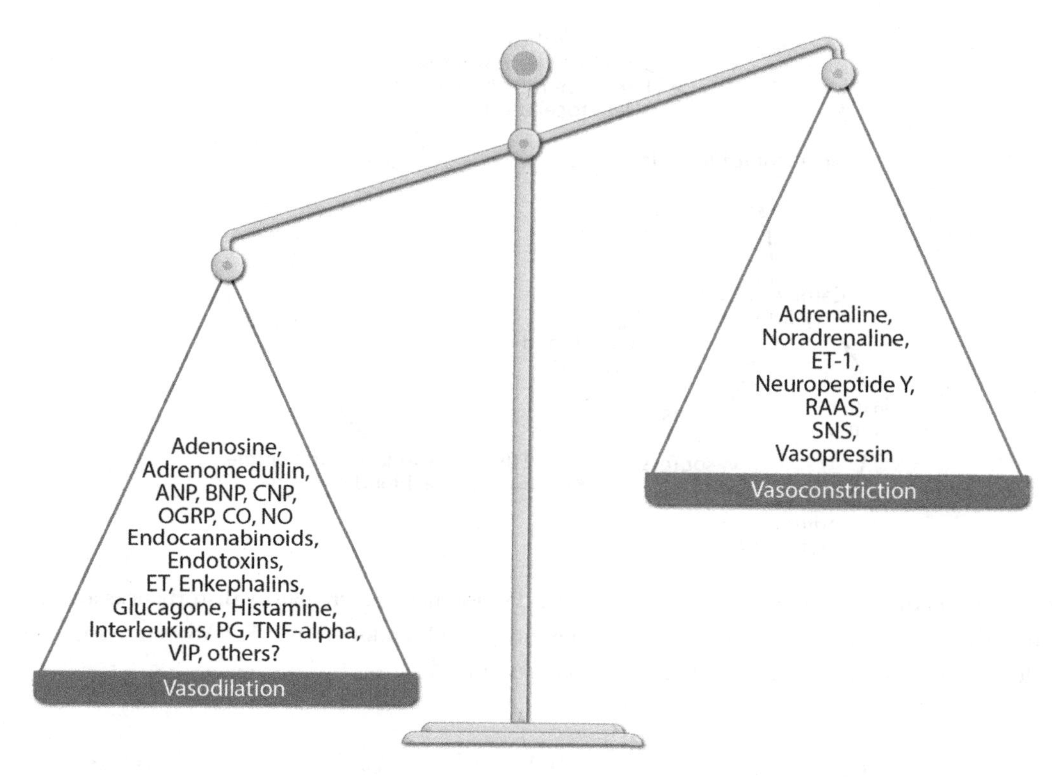

Figure 3.4a: Balance between vasodilators and vasoconstrictors in decompensated cirrhosis with ascites.

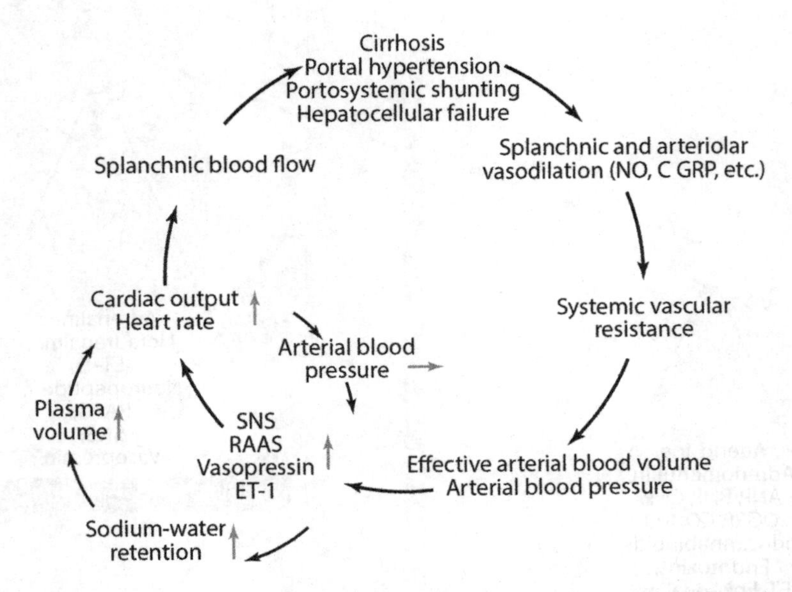

Figure 3.4b: Pathophysiological mechanisms in the development of circulatory disturbances and ascites in cirrhosis. CGRP: calcitonin gene-related peptide; ET-1: endothelin-1; RAAS: renin-angiotensin-aldosterone system; SNS: sympathetic nervous system.

3.1.2 DYNAMIC COUPLING BETWEEN THE HEART AND CENTRAL ARTERIAL TREE

Several recent studies in patients with heart failure and arteriosclerotic heart disease have indicated that the coupling between the left ventricle and the central arterial tree (ascending aorta and aortic arch) is very important for the load of the left ventricle and thereby the cardiac performance, fluid dynamics, and neurohumeral regulation [163, 164]. In patients with cirrhosis, especially in those with advanced disease, the arterial afterload is reduced. This is brought about by an overall reduction in the systemic vascular resistance. However, clinical investigations have also shown that the arterial compliance is substantially increased [9, 11, 147]. This is caused by both structural and functional alterations of the arteries [165] and also by the low or low normal arterial blood pressure in these patients [166]. Arterial pulse wave velocity is reduced and the back reflections of the arterial pulse are delayed [147]. This provides protection against strain to the left ventricle, and recent investigations of the arterial pulse in cirrhosis by fast Fourier analysis have confirmed this [167, 168]. Bringing the arterial blood pressure back to normal level with a vasoconstrictor like terlipressin increases the afterload and pulse velocity [169, 170]. This can unmask a latent left ventricular failure, and recent studies of left ventricular performance, wall motion, and wall thickness have indicated that normal arterial blood pressure and normal arterial compliance (i.e., normal afterload) unmask a latent left ventricular dysfunction [158, 171]. The presence of latent and manifest cardiac dysfunction and failure will aggravate the effective central arterial underfilling, with further activation of neurohumeral counterregulatory mechanisms, and aggravate renal dysfunction with avid sodium water retention and upcome of a hepatorenal syndrome at the end stage [142, 172-176].

3.2 NEUROHUMORAL REGULATION IN HEPATIC ASCITES

Sinusoidal and portal hypertension in cirrhosis is related to a defective nitric oxide production [114, 132, 177, 178]. Endogenous vasoconstrictors like endothelin 1, angiotensin II, and catecholamines may also increase the sinusoidal vascular resistance in addition to structural changes [114, 141, 178, 179]. However, the coupling between portal-sinusoidal hypertension and the genesis and perpetuation of systemic and splanchnic vasodilatation is not completely understood [9, 85]. It may be caused either by direct neurohumoral signals from the liver [131] or by an overproduction of circulating vasodilators induced by shear stress. Several findings suggest that the splanchnic vasodilatation precedes renal sodium and water retention [160, 161, 178]. In experimental and clinical portal hypertension, splanchnic vasodilatation leads to reduced systemic vascular resistance, decreased effective arterial blood volume, and a reduction in arterial blood pressure [144, 147]. This brings about activation of potent systemic vasoconstrictor systems [14], and the hemodynamic consequences include a hyperdynamic systemic circulation [11, 144, 154]. Recently, it became clear that in advanced cirrhosis further underfilling of the arterial circulation is also secondary to a reduction of the increased cardiac output as described in patients with renal failure and spontaneous bacterial peritonitis [159, 174, 175, 180-182].

The highly activated sympathetic nervous system, RAAS, and increased circulating vasopressin contribute to sodium-water retention, extracellular fluid volume dysregulation, and formation of oedema and ascites. However, these systems are not only highly activated, they are also dysfunctional [138, 172, 183-185]. Thus, a down-regulation of beta-adrenergic receptors has been described [186-188]. Parasympathetic and sympathetic dysfunction is present at several levels: central nervous system, peripheral nerves, pre-synaptic, synaptic, and post-synaptic [189, 190].

In early cirrhosis the RAAS is normal or slightly activated [138], but in some patients it may be suppressed, suggesting a primary sodium-water retention [21]. One should remember that normal plasma renin activity in a patient with expanded plasma volume indicates RAAS overactivity [13, 14, 191]; see Table 3.1. In advanced cirrhosis, high activation of the RAAS is the typical finding [138]. Reduced sensitivity to aldosterone in the renal tubules has been described [192-194], and smooth muscle cells from patients with cirrhosis and animals with experimental cirrhosis have a reduced sensitivity to noradrenaline, angiotensin II and vasopressin [114]. A characteristic feature of human and experimental cirrhosis is lack of mineralocorticoid escape [21, 195].

Table 3.1: Evidence of reduced central and effective arterial filling in patients with cirrhosis

	Cirrhosis	Controls
Central and arterial blood volume relative to blood volume (fraction, CBV/BV)	0.25±0.04* (n=60)	0.34±0.10 (n=22)
Central and arterial blood volume relative to cardiac output (sec, CBV/CO)	11.8±2.9* (n=60)	19.4±4.4 (n=22)
Plasma volume relative to plasma renin activity (10^{-15} ml²/pg², PV/PRA)	3.41* (n=89)	7.6 (n=32)

Mean ± SD. *Cirrhosis vs controls: $p < 0.001$.
In most clinical cases an increased plasma renin activity (PRA) will be taken as evidence of reduced arterial filling in patients with cirrhosis.
CBV: central and arterial blood volume; BV:blood volume; PV:plasma volume; CO:cardiac output.
Data from Henriksen et al. [106, 149].

Nitric oxide is an important vasodilator in the systemic vasodilatation [177, 178, 196], with an up-regulation of the nitric oxide synthesis [114, 178]. Calcitonin gene-related peptide (CGRP) and adrenomedullin are potent neuropeptide vasodilators that are increased, especially in advanced cirrhosis with ascites and the hepatorenal syndrome [100, 197-200]. Systemic vasodilatation has also been related to resistance to vasopressors. An impaired response to vasoconstrictors is related to changes in receptor affinity, down-regulation of receptors and post-receptor defects that may be related to increased expression of nitric oxide [114, 178].

The pathophysiology and implications of arterial vasodilatation are complex. Definite experimental and clinical evidence shows that it precedes the activation of counterregulatory neurohumeral activation and the renal sodium and water retention in cirrhosis. Therefore it plays a primary and major role in both local control of extracellular fluid volume dysregulation and overall sodium-water dynamics in cirrhosis.

3.3 KIDNEY FUNCTION IN ASCITES WITH SPECIAL FOCUS ON HEPATIC ASCITES

Characteristic changes in renal function follow hepatic insufficiency, especially in patients with cirrhosis. It should be noted that owing to altered metabolism, body composition and dietary sodium intake, and excess alcoholic intake, some adaptive changes in renal function may also take place in patients with cirrhosis [201, 202]. By substituting carbohydrates and lipids by isocaloric ethanol, rats show a characteristic reduction in glomerular filtration rate (GFR) with a reduced area of the glomerular membrane [203]. Conversely, increased protein intake will increase GFR, and reduced protein intake will reduce GFR [204]. In the aging person, GFR decreases concordantly with the reduction in skeletal muscle mass leaving the serum creatinine concentration almost unchanged. A long-term reduction in the sodium intake may also reduce RBF and GFR. Thus, in the cirrhotic patient with a high ethanol intake, low intake of protein and sodium, reduced skeletal muscle mass, and intake of diuretics, reduced kidney function would be expected for purely adaptive reasons [22, 205, 206].

Renal dysfunction with relation to extracellular volume regulation is considered in the following. Renal sodium and water retention in advanced cirrhosis is the most avid seen in human pathophysiology [140]. A 24 h sodium excretion of only one mmol or even less is not exceptional. However, it must be realized that all parts of the nephron exhibit altered physiology, and it is in the main a functional type without structural changes [180, 207]. The characteristic renal dysfunction here is a reduced renal blood flow (RBF), reduced GFR, increased proximal tubular sodium reabsorption, increased distal tubular sodium reabsorption, and reduced urinary water excretion, owing to enhanced distal tubular and collecting duct reabsorption of water [144, 208, 209].

The changes in cirrhosis may be pronounced and constitute a terminal stage in advanced disease with functional renal failure, either progressing quickly as in the hepatorenal syndrome type 1 (HRS-1), or more slowly in the hepatorenal syndrome type 2 (HRS-2) [22, 23, 142, 172, 174].

3.3.1 RENAL BLOOD FLOW (RBF)

In certain types of experimental cirrhosis (CCl_4 in rats, dimethylnitrosamine in dogs), the RBF may be unchanged or even increased [140, 210]. Increased RBF and hyperfiltration have also been described in a few patients with early pre-ascitic cirrhosis [209, 211]. However, at present there is no general agreement as to the fraction of cirrhotic patients with this phenomenon [15, 88, 138, 212, 213]. A characteristic change in moderate and advanced cirrhosis is a reduction in RBF [208]; see Figure 3.5. This is brought about by constriction in both the afferent and the efferent arteriole [207]. In early cirrhosis the efferent arteriole may be more constricted than the afferent, thus reducing perfusion more than filtration (increased filtration fraction). Enhanced sympathetic nervous activity, elicited by effective arterial hypovolaemia, reduces the RBF, owing to alpha-adrenergic stimulation [106, 214]. Similarly, an inverse relation between renal venous noradrenaline and RBF and between the central blood volume and renal venous noradrenaline has been described [207]. The beta-adrenoceptors are also stimulated, which give rise to increased renin production with elevated angiotensin II, which also reduces the renal perfusion.

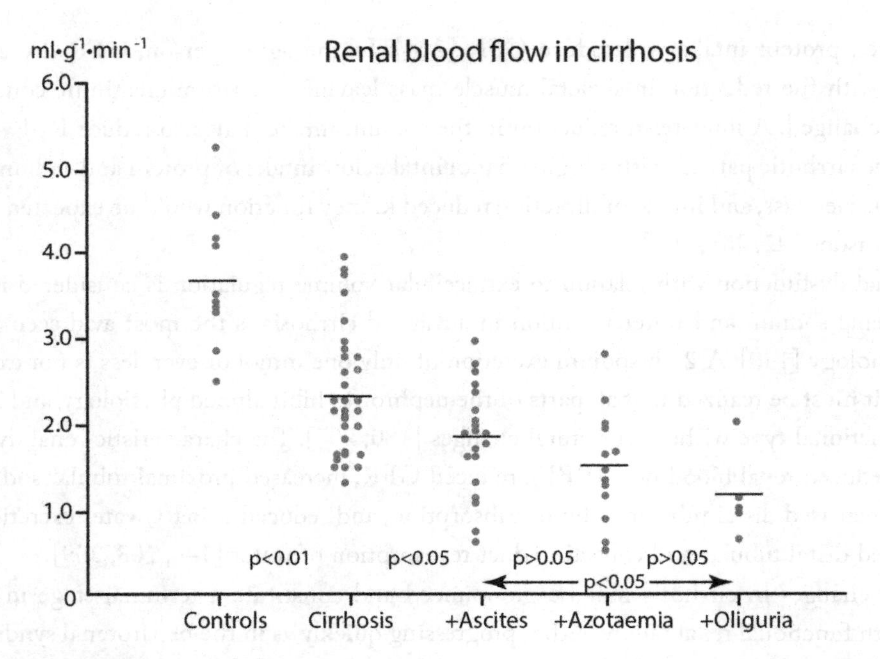

Figure 3.5: Renal blood flow (Y-axes) in controls, compensated cirrhosis, decompensated cirrhosis, and patients with hepatorenal syndrome. There is a progressive fall in renal blood flow with the severity of liver disease. From Ring-Larsen [208].

Circulating endothelin 1, a very powerful peptide vasoconstrictor of renal and coronary arteries, may be increased and contribute to the vasoconstriction seen in renal dysfunction and the hepatorenal syndrome [141]. Thus, Moore and co-workers reported a higher concentration of endothelin 1 in the renal vein compared to the artery in patients with the hepatorenal syndrome [215]. Moreover, endogenous renal vasodilators like the prostaglandins are reduced in patients with cirrhosis [216]. In addition, it has recently been shown that calcium/polyvalent cation receptors may be down-regulated in the smooth muscle of renal vessels, with a potentially increased renal vasoconstriction as the outcome [217].

Another mechanism of reduced RBF is the lowering of the arterial blood pressure and the reduced effective renal perfusion pressure, owing to increased renal venous pressure in some cirrhotic patients, because of the presence of ascites and elevated pressure in the inferior vena cava [8, 143, 218]. The effective renal perfusion pressure may thus be reduced from 85 mmHg in the normal condition to 50 mmHg or even less in advanced cirrhosis.

The normal response to activation of renal sympathetic nerves is increased renin secretion, increased proximal tubular reabsorption of sodium, and with intensified sympathetic activity the RBF and GFR will decrease [140]. Renal hypoperfusion is, at least initially, a physiological response to changes in the systemic circulation with effective arterial hypovolaemia [218]. The increased activity of plasma renin and plasma noradrenaline concentrations correlates inversely with the reduction in the RBF and GFR [106]. As angiotensin II chiefly acts on the efferent arteriole, ACE inhibition may induce a significant reduction in the GFR and filtration fraction and further reduction in sodium excretion, even in the absence of any change in arterial blood pressure or RBF [218].

3.3.2 GLOMERULAR FILTRATION RATE (GFR)

GFR is governed by a balance in the tonus of the afferent and efferent arterioles of the nephron. Thus, if the afferent arteriole is constricted, the GFR may be relatively more affected than RBF. When a simultaneously high tone is present in the efferent arteriole, the filtration pressure will be maintained [131]. In such a case, the filtration fraction will be high, as found in early ascitic patients and patients without the hepatorenal syndrome [207, 219]. Severely decompensated patients and patients with the hepatorenal syndrome have a very high sympathetic nervous activity. These patients are characterized by a very low RBF, especially in the cortical area, substantially decreased GFR, and low filtration fraction, indicating preferential constriction of the afferent arterioles [140, 208]. In advanced decompensated cirrhosis, the kidneys appear to function like bilateral Goldblatt kidneys [140]. Keeping up the arterial blood pressure results, as judged from earlier experiments with ornipressin [208] or more recent experiments with terlipressin [153, 170, 220], in a substantial increase in GRF and reduction in serum creatinine.

3.3.3 PROXIMAL TUBULES

In the normal condition, about 80% of filtered sodium is reabsorbed in the proximal tubules and the preurine is isotonic when it leaves the proximal tubules [131]. In cirrhosis, the proximal tubular reabsorption fraction (i.e., sodium reabsorption as a fraction of the filtered sodium load) is normal or increased [142, 193, 221], which indicates that the distal sodium delivery is reduced [153, 209, 219, 222]. The increase in proximal sodium reabsorption is in part brought about by enhanced alpha-adrenergic receptor activity [214]. Various animal models of cirrhosis (CCl_4, bile duct ligation in rats) may give somewhat different results (140). Particularly in experimental CCl_4 cirrhosis in rats, hyperfiltration has been reported with high sodium reabsorption in proximal and distal tubules [223]. However, in advanced human cirrhosis with diuretic-resistant or diuretic-intractable ascites, failure of natriuresis is often bound to enhanced proximal sodium reabsorption, which may sometimes be overcome by the supra-addictive effect of a thiazide diuretic [207, 218].

3.3.4 THICK ASCENDING LIMB OF HENLE'S LOOP AND DISTAL TUBULES

Increased distal sodium reabsorption has been described in patients with cirrhosis and animals with experimental cirrhosis [140, 224]. This has been attributed to the effect of increased circulating aldosterone [209, 225], increased sensitivity of the distal tubules to aldosterone [193, 226], down-regulation of the enzyme 11-beta-hydroxy-steroid dehydrogenase with inappropriate activation of tubular sodium reabsorption by endogenous glucocorticoids [227], or down-regulation of renal calcium/polyvalent cation sensing receptors with up-regulation of tubular sodium retaining channels [217, 228]. Most likely a combination of different mechanisms exists, especially in advanced cirrhosis [229].

3.3.5 COLLECTING DUCTS

Stimulation of vasopressin-2 receptors (V2 receptors) activates adenylate cyclase [3, 230]. Increased intracellular cyclic adenosine monophosphate (cAMP) increases the water transport in the collecting ducts due to an increased number of aquaporin 2 water channels (AQP2) in the apical cellular membrane [231]. The short-term up-regulation involves increased intracellular recycles of AQP2 molecules to the apical membrane; see Figure 3.6. Increased intracellular cAMP stimulates also genomic transcription in the long-term up-regulation of AQP2. Thus, by stimulation of V2 receptors, endogenous arginine-vasopressin (AVP) increases the number of AQP2 and thereby enhances collecting duct reabsorption of water with reduced free water clearance as the outcome [232-234].

In patients with cirrhosis, there is a non-osmotic stimulation of neuropituitary release of AVP with increased stimulation of collecting duct V2 receptors and enhanced transport of water from the urine back to the plasma, reducing the free water clearance, even in the presence of hyponatremia [230, 235-237]. Although most of the APQ2 is recycled and endocytosed into ductal epithelial vesicles, a small part spills over into the urine, where the content of APQ2 may reflect the stimulation of V2 receptors and thereby the increased action of AVP on the collecting ducts [234, 237-239].

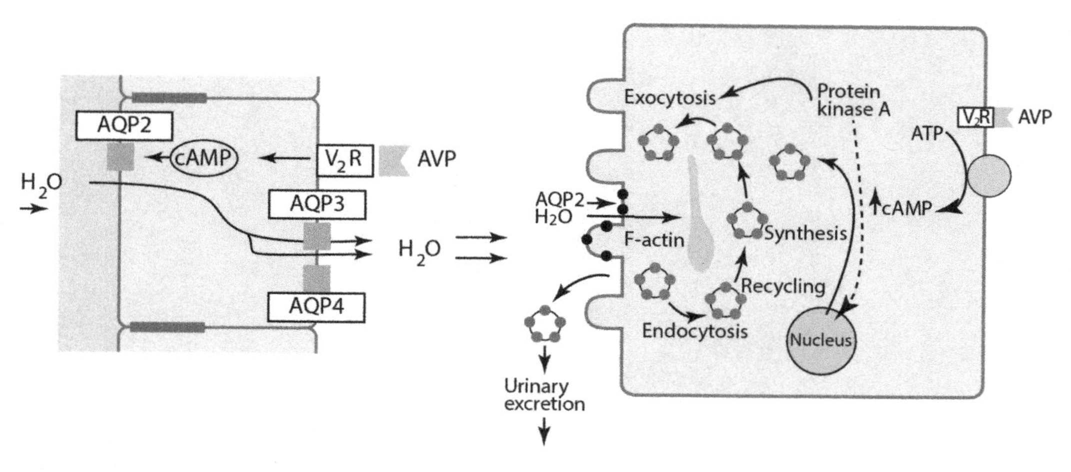

Figure 3.6: Regulation of water permeability in collecting ducts. Arginine-vasopressin (AVP) has affinity for the V2 receptor (V2R) and increases intracellular cyclic AMP (cAMP), which stimulates the phosphorylation of the aquaporin molecule. Especially, Aquaporin-2 (AQP2) is then translocated from the intracellular vesicles to the apical membrane. Water permeability increases and facilitates the osmotically driven transport of water from the tubular lumen into the interstitium. From Krag et al. [230].

CHAPTER 4
Complications of Ascites

4.1 HEPATIC NEPHROPATHY AND THE HEPATORENAL SYNDROMES

In the early stage of cirrhosis, impairment of renal function is evidenced by reduced renal sodium excretion after acute administration of a sodium chloride load or change in body position [140]. Later, reduction in the RBF and GFR with increased renal sodium and water reabsorption in parallel with the reduction in liver function takes place [207, 208]. In addition these patients have reduced free water clearance and may develop reduced serum sodium or a full-blown dilutional hyponatraemia with a serum sodium concentration below 125 mmol/l [7, 153, 237, 240, 241]. In the end-stage, progressive reduction in the RBF and GFR leads to development of the hepatorenal syndrome (HRS). Two types of HRS have been defined, depending on how fast the renal failure progresses [22, 23, 201]. Type 1 HRS has a rapid course whereas the progression of type 2 HRS is more slow and related to the presence of refractory ascites [23,159]. In particular the prognosis of type 1 HRS has remained poor [174, 201; see Figure 4.1]. The effective arterial hypovolaemia and the massive activation of vasoconstrictor systems are central elements in the pathogenesis of HRS [8, 201, 207, 242]. A prerequisite for the development of HRS is advanced liver disease with severely disturbed liver function and sinusoidal-portal hypertension. Normalization of renal function and regulation of extracellular fluid volume after orthotopic liver transplantation may indicate that the liver is directly involved in the disturbance of renal function. The existence of a hepatorenal reflex has been described in animal studies, but has been debated for years in man [214, 243-245]. The presence of a hepatorenal reflex in man is supported by observations of reduced RBF following a rise in portal pressure and an increase in the renal release of endothelin 1 [215, 243].

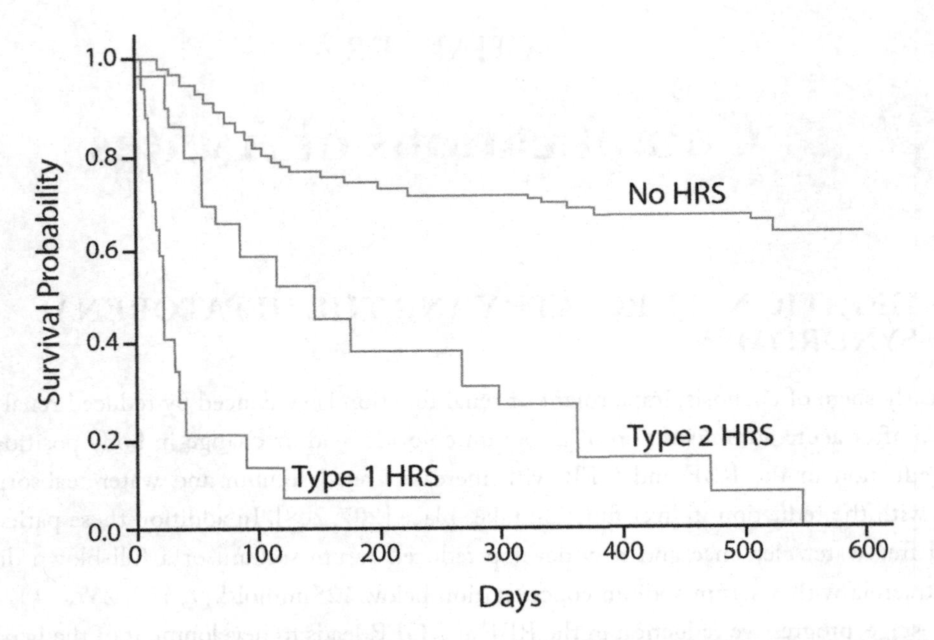

Figure 4.1: Survival in patients no hepatorenal syndrome (HRS), type 2 HRS, and type 1 HRS, respectively. From Arroyo et al. [174].

In healthy subjects the renal autoregulation maintains normal renal perfusion in spite of alterations in arterial blood pressure, provided it is above approximately 70 mmHg; see Figure 4.2. Below this threshold RBF is directly related to the renal perfusion pressure, which equals the arterial mean pressure as long as the renal venous pressure is low [131, 246]. In animals with increased sympathetic nervous activity or infusion of sympathomimetic agents, the renal autoregulation curve shifts towards the right side [131]. It is well known that enhanced renal sympathetic nervous activity in man will also shift the renal autoregulation curve to the right [214, 246]. In patients with cirrhosis, there are indications that the relation between RBF and renal perfusion pressure shifts to the right with a lower RBF, also in the case of normal renal perfusion pressure [207, 218]; see Figures 3.4a, 3.4b and 4.2. This means that even with a low normal or normal arterial blood pressure, patients with cirrhosis may be very sensitive to a further reduction in the mean arterial blood pressure and thereby renal perfusion pressure. This is especially true in the presence of ascites, where renal venous pressure is elevated [143]. In the presence of cirrhotic cardiomyopathy, with diastolic and systolic dysfunction, and spontaneous bacterial peritonitis (both of which are frequently associated with HRS, see later) the low arterial blood pressure may further decrease the renal perfusion [159, 174, 175]. In patients with advanced liver disease the arterial blood pressure and effective renal perfusion pressure may be so low that elevation of the arterial blood pressure with vasopressin (terlipressin) or another vasoconstrictor may increase the RBF and thereby improve renal function, simply by moving up the first part of the renal autoregulation curve [140, 153, 218, 220, 247, 248]. Prevention of effective arterial hypovolemia, arterial hypertension, and cardial dysfunction are therefore important targets for therapy of ascites and prevention of the HRS.

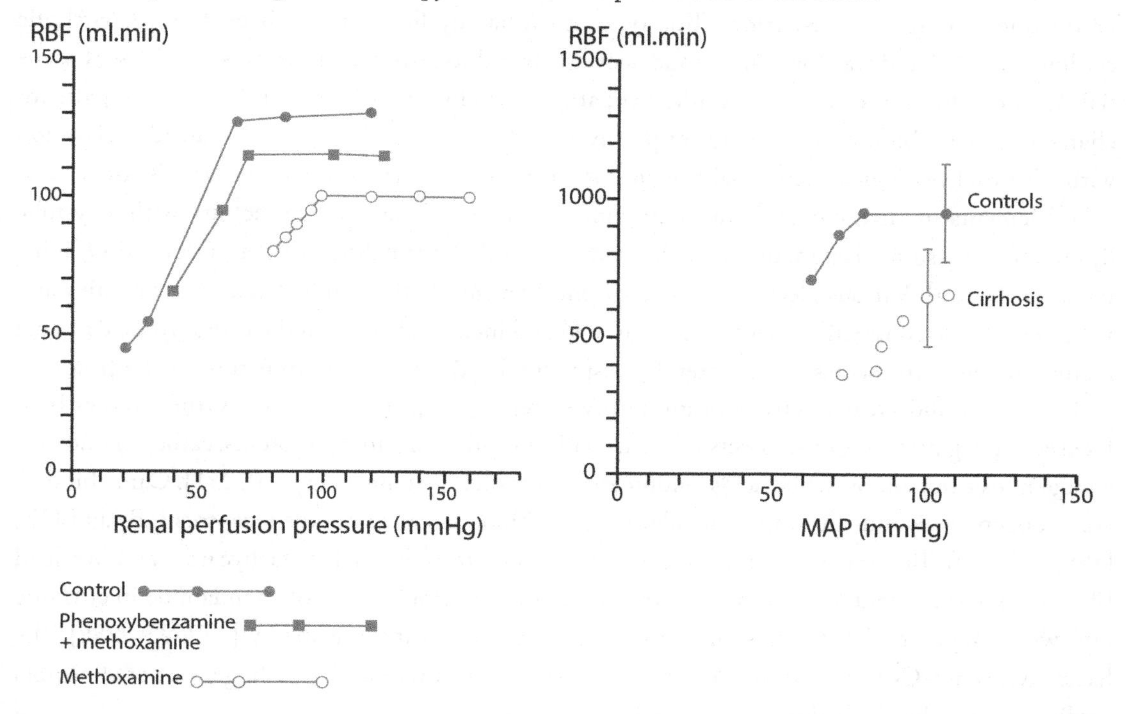

Figure 4.2: Autoregulation of renal perfusion. X-axis shows renal perfusion pressure (mean arterial blood pressure minus renal venous pressure), Y-axis shows renal blood flow (RBF). The curves illustrate control subjects and patients with cirrhosis. It is seen that the autoregulation curve in cirrhosis is shifted to the right and that the level of RBF is lower in patients with cirrhosis compared to the controls.

The GFR is reduced below 30 ml/min in patients with the HRS. The amount of sodium reaching the distal nephron is reduced and may explain why diuretics such as furosemide and spironolactone may be of limited value in these patients [218, 223]. The low free water clearance consequent on massive V2 receptor stimulation leads to dilutional hyponatraemia [7, 159]. The use of V2 receptor antagonists and opioid antagonists are important issues for future research and potential therapy of dilutional hyponatraemia [174, 236, 249].

4.2 CARDIAC DYSFUNCTION IN HEPATIC ASCITES

The expanded blood volume in advanced cirrhosis contributes to a persistent increase in cardiac output, which may overload the heart [150]. Under other circumstances, increased cardiac output and augmented cardiac work would cause cardiac failure, but because of the decreased afterload, as reflected by reduced systemic vascular resistance and increased arterial compliance, left ventricular failure may be latent in cirrhosis [9, 158, 189]. Cardiac failure may become manifest under strain or treatment with vasoconstrictors. This type of cardiac dysfunction has been termed "cirrhotic cardiomyopathy" and was for years erroneously attributed to alcoholic heart muscle disease. Essentials in the definition of cirrhotic cardiomyopathy are a chronic dysfunction in cirrhotic patients, characterized by blunted contractile responsiveness to stress, and/or altered diastolic relaxation with electrophysiological abnormalities in the absence of other known cardiac disease (Table 4.1). Elements in cirrhotic cardiomyopathy include impaired cardiac contractility with a systolic dysfunction, diastolic dysfunction and electromechanical abnormalities with a prolonged Q-T interval [158, 189]. Various electrophysiological mechanisms of the conductance abnormalities and impaired cardiac contractility have been suggested and include changes in the cadiomyocyte plasma membrane with an increased cholesterol/phospholipid ratio, attenuated function of the β-adrenergic pathway and greater activity of inhibitory systems [189]; see Figure 4.3. Other studies have focused on negative inotropic effects of nitric oxide, nitration of cardiac proteins, carbon monoxide, endogenous cannabinoids, bile acids, endotoxins, and other systems [158, 250, 251]. Cannabinoids are endogenous ligands including anandamide that binds to cannabinoids receptors CB_1 and CB_2 [189, 252-254]. The production may increase in response to stress such as tachycardia and overload [254, 255]. Experimental studies have shown a negative inotropic effect of anandamide in cirrhotic rats, which suggests that this system is involved in cirrhotic cardiomyopathy [189, 256-258]. The haem oxygenase-CO pathway has also been shown to play a role in the pathogenesis of abnormal cardiac contractility in cirrhotic cardiomyopathy [158].

Table 4.1: Diagnostic tools and criteria of cirrhotic cardiomyopathy
• Present in the face of a hyperkinetic circulation with a combined systolic and diastolic dysfunction together with electrophysiological abnormalities. • Different from alcoholic heart muscle disease • Systolic dysfunction demasked by physical or pharmacological stress • Diastolic dysfunction detected by echocardiographic measurement of the E/A ratio • Q-T interval prolongation assessed on the ECG and adjusted adequately
Systolic function
Echocardiografphy/MRI: • Volumes • Fractional shortening • Velocity of fractional shortening • Ejection fraction • Response to stress (dobutamine) • Wall motion • Tissue doppler amd speckle tracking
Exercise ECG: • Exercise capacity • Oxygen consumption • Pressure x heart rate product
Myocardial perfusion imaging with gating: • Regional myocardial perfusion • Cardiac volumes • Ejection fraction • Wall motion and wall thickening
Diastolic function
Echocardiography/MRI/radionucleid angiography: • E/A ratio • Deceleration time • A and E waves • Relaxation times
E/A ratio, ratio of early to late (atrial) phases of ventricular filling.

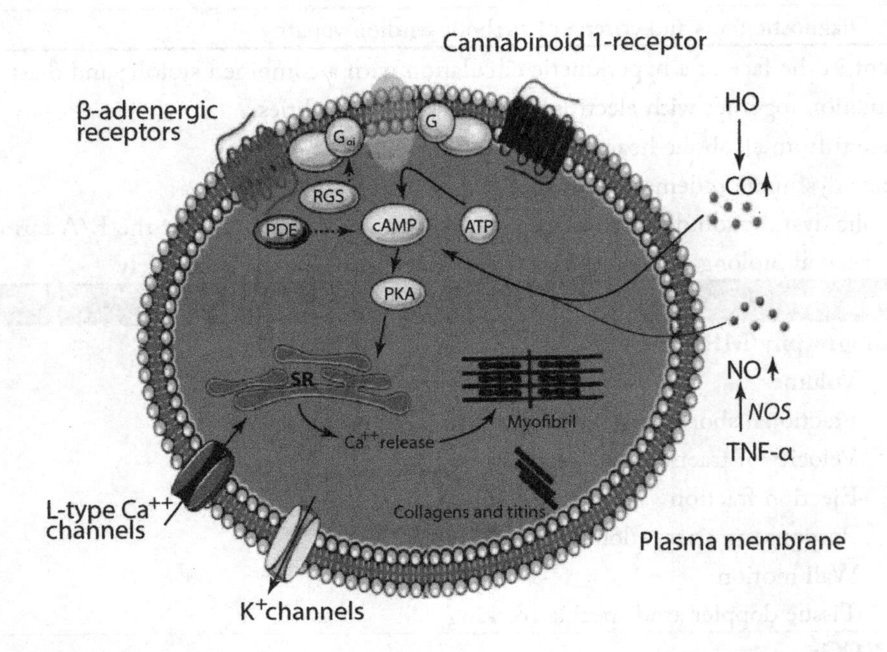

Figure 4.3: Altered cardiomyocytes in cirrhotic cardiomyopathy. Nitric oxide (NO) and carbonmonoxyde (CO) change intracellular signal transduction. There is a down regulation of beta-adrenergic receptors, a change in plasma membrane fluidity and activation of cannabinoid 1-receptors. The cirrhotic cardiomyopathy syndrome includes altered conduction in potassium channels and in certain types of calcium channels as well as calcium release from the sarcoplasmatic reticulum (SR). From Møller et al. [158].

4.2.1 SYSTOLIC DYSFUNCTION

In cirrhotic cardiomyopathy, the left ventricular end-diastolic pressure increases after exercise, but the expected increases in cardiac stroke index and left ventricular ejection fraction (LVEF) are absent or subnormal, which indicates an inadequate response of the ventricular reserve to a rise in ventricular filling pressure [259]. A vasoconstrictor-induced increase of 30% in the left ventricular afterload results in an approximate doubling in pulmonary capillary wedge pressure, with no change in cardiac output [260]. Recently, Møller and coworkers have shown by myocardial perfusion imaging that infusion of terlipressin suppresses myocardial function, whereas the myocardial perfusion is left unaffected [261]. This response may be useful in diagnosing cirrhotic cardiomyopathy. A similar pattern is seen after insertion of a transjugular intrahepatic portosystemic shunts (TIPS), but the raised cardiac pressures after TIPS tend to normalise with time [262, 263]. Some of these patients (12%) may develop manifest cardiac failure in association with the TIPS insertion [264]. Similar effects are seen after infusion of plasma expanders. Infusion of a plasma protein solution, however, increases cardiac output, as well as right atrial pressure, pulmonary arterial pressure, and pulmonary capillary wedge pressure, whereas infusion of packed red blood cells may not produce a change in these variables [11, 144].

The LVEF reflects systolic function, even though it is very much influenced by preload and afterload. It has been reported to be normal at rest in some studies and reduced in one study of a subgroup of patients with ascites [259, 265, 266]. After exercise, LVEF increases less in cirrhotic patients than in controls [158, 259, 267]. The reduced functional capacity may be attributed to a combination of blunted heart rate response to exercise, reduced myocardial reserve, and profound skeletal muscle wasting with impaired oxygen extraction [268, 269]. In patients with advanced cirrhosis and ascites, severe vasodilatation, activation of the RAAS, impaired renal function, and a reduced systolic function (a decrease in cardiac output) appear to be major determinants for the development of the hepatorenal syndrome [180]. Spontaneous bacterial peritonitis is a well-known risk factor for the development of the hepatorenal syndrome, and after resolution of the infection, suppression of systolic function appears to be more pronounced in patients who develop renal failure. Maintenance of cardiac contractility thus appears to be an important factor in the prevention of renal failure [135, 182].

4.2.2 DIASTOLIC DYSFUNCTION

Many patients with cirrhosis exhibit various degrees of diastolic dysfunction, which implies changes in myocardial properties that affect left ventricular filling. Diastolic dysfunction may progress to systolic dysfunction, although this has not been directly shown in cirrhotic patients [158, 270]. The pathological basis of the increased stiffness of the left ventricle seems to be cardiac hypertrophy, patchy fibrosis and subendothelial oedema [158, 267, 271, 272]. Determinants of a diastolic dysfunction on a Doppler echocardiogram are decreased E/A ratio (the ratio of early to late (atrial) phases of ventricular filling) and delayed early diastolic transmitral filling with prolonged deceleration and isovolumetric relaxation times (Table 4.1) [158, 266, 273]. In a number of studies, A wave and E wave velocities and deceleration times are much increased and E/A ratio is decreased in cirrhotic patients, especially in those with ascites [266, 270, 274]. Recent studies of ventricular diastolic filling in cirrhosis support the presence of a subclinical myocardial disease with diastolic dysfunction, which, in ascitic patients, improves after paracentesis and can be aggravated with TIPS [262, 266, 275, 276]. In these decompensated patients, paracentesis seems to ameliorate diastolic, but not systolic, function [266]. Patients with TIPS with an E/A ratio <1 seem to have a poorer survival rate than patients without signs of diastolic dysfunction [275]. Liver transplantation has recently been shown to reverse cardiac changes including diastolic dysfunction [267, 277]. It has been proposed that diastolic dysfunction precedes systolic dysfunction in early heart disease and that anti-aldosterone treatment improves cardiac function. Pozzi et al. recently demonstrated that anti-aldosterone treatment with K-Canrenoate in cirrhosis ameliorated cardiac structure by reducing left ventricular wall thickness and volume, but had almost no effects on systolic and diastolic functions [278]. It is also possible that anti-aldosterone treatment may have beneficial effects on catecholamine-induced cardiac fibrosis, as described in heart failure [279].

The clinical significance of diastolic dysfunction and its importance in cirrhotic cardiomyopathy has been questioned, as overt cardiac failure is not a prominent feature of cirrhosis. However, there are several reports of unexpected death from heart failure following liver transplantation, surgical portocaval shunts, and TIPS [264, 277, 280]. These procedures involve a rapid increase in cardiac preload. In a less compliant heart, the diastolic dysfunction could be enough to cause pulmonary oedema and heart failure. This is consistent with the findings of Huonker et al. [262], who reported an increase in pulmonary artery pressure, preload, and diastolic dysfunction after TIPS. In patients with the hepatopulmonary syndrome and in children with chronic hepatitis, an isolated right ventricular diastolic dysfunction has been described and it may play a role in the right cardiac function and clinical course of these patients [281, 282]. Thus, both left and right diastolic dysfunction could account for part of the cardiac dysfunction in cirrhotic cardiomyopathy in patients with ascites.

4.2.3 ELECTROMECHANICAL ABNORMALITIES

There is a large body of evidence for electrophysiological abnormalities in cirrhosis, especially in patients with ascites, primarily comprising prolonged repolarization time and increased dispersion of the electromechanical time interval [283-285]. The sympathetic nervous activity influences the heart rate and electromechanical coupling by several mechanisms: noradrenaline binding to β-receptors, receptor-mediated G protein interaction and, consequently, stimulation of adenylcyclase, activation of cAMP-dependent phosphokinase A and channel phosphorylation; see Figure 4.3. Several receptors and postreceptor defects have been described in cirrhosis with reduced β-receptor density and sensitivity, and altered G protein and calcium channel functions [158, 271, 286]. All these defects may explain both impaired chronotropic responses and electromechanical uncoupling. The coupling between cardiac contractions and the arterial system is of a major importance for the amount of work performed by the left ventricular myocardium, and thereby for the strain on the heart [268, 287]. In addition, Ward et al. showed a decrease in K^+ currents in ventricular cardiomyocytes from cirrhotic rats, which prolongs the Q-T interval [288]. The prolonged repolarization time is reflected by a prolonged Q-T interval in a substantial fraction of the patients with advanced cirrhosis, which could lead to ventricular arrhythmias and sudden cardiac death, but the evidence from clinical studies is sparse [158, 271]. In cirrhotic patients, the prolonged Q-T interval is significantly related to the severity of the liver disease, portal hypertension, presence of ascites, portosystemic shunts, elevated brain-type natriuretic peptide (BNP) and pro-BNP, elevated plasma noradrenaline, and reduced survival [103, 284, 289, 290]. The prolongation of the Q-T interval is partly reversible after liver transplantation and β-blocker treatment [45,82]. The prolonged Q-T interval in cirrhosis should be considered an element in the cirrhotic cardiomyopathy and may be of potential use in identifying patients at risk [285]. Conclusively, impaired cardiac function in ascites may be both primary and secundary to the genesis and perpetuation to the ascites syndromes. Evaluation of cardiac function is important in all patients with ascites.

4.3 OTHER COMPLICATIONS AND EFFECTS OF ASCITES ON THE ORGANISM

4.3.1 PHYSICAL EFFECTS OF ASCITES

Slight ascites does not give any appreciable discomfort apart from cosmetic nuisance. *Moderate* and *tense ascites*, patients in whom the intraabdominal pressure may rise 10-20 mmHg above that of the right atrium [5, 291], may cause abdominal pain and respiratory distress (Figure 2.3). In addition, an adverse effect on cardiovascular functions may exist in patients with tense ascites, as indicated by studies by Guazzi and co-workers [291].

4.3.2 UMBILICAL, INGUINAL AND OESOPHAGEAL HERNIA

Hernias may arise secondary to fluid accumulation in the peritoneal cavity, and they are not rare (about 25%). Rupture of an umbilical hernia with instantaneous voiding of the entire amount of ascitic fluid is infrequent in patients with ascites, but is generally considered a bad prognostic sign [88].

4.3.3 SUBACUTE BACTERIAL PERITONITIS (SBP)

SBP is defined as the presence of a neutrophil cell count above 250 per μl ascitic fluid. A positive cell culture with growth of most often only one microbiological agent is characteristic. If more than two microbes are present, or in case of a very high neutrophil cell count, secondary bacterial peritonitis should be suspected [88].

Patients with ascites exhibit a variety of immunologic incompetence with an increased risk of bacterial infections, and development of SBP [292, 293]. Typically, patients with SBP do not initially present symptoms of infection but some patients may show signs of shock, hepatic encephalopathy, or deterioration of the liver function. The mortality of SBP without treatment is above 50%, but it can be reduced to less than 20% with early diagnosis and specific treatment [122, 294, 295]. SBP is seen in approximately 10% of the patients who are hospitalized with cirrhosis and ascites [88, 122, 296]. Newly hospitalized patients with ascites should therefore immediately undergo a diagnostic paracentesis. SBP is distinguished from secondary bacterial peritonitis by the absence of an intraabdominal focal infection. Diagnostic paracentesis should be repeated during hospitalization by suspicion of any infection especially with deterioration of liver, heart, or kidney function [297].

Cell count performed with automatized equipments such as a flow cytometer seems to be as effective as microscopy. Cell analysis should be accessible 24 h a day with a response time not exceeding six hours. The results of culture of ascitic fluid should also be available immediately [88, 296].

4.3.4 HEPATORENAL SYNDROME (HRS) AND CIRRHOTIC CARDIOMYOPATHY

HRS and cirrhotic cardiomyopathy are considered elsewhere.

CHAPTER 5

Treatment of Ascites

5.1 TREATMENT OF HEPATIC ASCITES

A short description of treatment principles is included in this book when it is relevant for the understanding of aetiology, pathophysiology, and clinical features of the ascitic syndrome. In general, treatment of patients with cirrhosis and portal hypertension should be based on knowledge of essential pathophysiological aspects and in particular the size of the portal pressure. This can only be achieved by measurement of the HVPG during a liver vein catheterization (see Figure 2.4). However, such facilities may not be available in all centers and in general practice where the treatment may be guided by clinical signs and biochemistry.

5.1.1 TREATMENT OF NON-COMPLICATED ASCITES

Previous studies have shown that the supine position ameliorates the reduced RBF and GFR and improves sodium and water excretion [298]. A less activated RAAS and SNS and a more favorable diuretic response in supine patients have led to the assumption that bed rest may benefit the treatment of ascites. However, severe side effects of prolonged bed rest, i.e., increased risk of thrombo-embolic complications, decalcification of bones, and muscular atrophy, imply that bed rest in general cannot be recommended as routine treatment of ascites [88, 212].

Reduced salt intake counteracts the positive sodium balance in patients with fluid retention and may create a negative sodium balance in some patients. Therefore, a low salt diet is part of the basic treatment of ascites. However, a rigorous low salt diet is often unacceptable to the patient and based on a general experience a non-salt-added diet of 80-120 mmol NaCl per day is often recommended. In patients with dilutional hyponatraemia, water restriction (1.0-1.5 l/day) has been recommended but the effectiveness of such a treatment to increase the serum sodium concentration more than 5 mmol/l is limited [7, 88, 299]. Moreover, water restriction does not improve serum sodium concentration, because the daily fluid intake cannot be reduced to less than one liter per day, in order to ensure a negative fluid balance [300]. Water restriction should therefore be reserved to patients who are clinically hypervolaemic with severe hyponatraemia, and decreased free water clearance. Thus, water restriction can only be applied clinically in very few patients.

Medical Treatment. A survey of medical treatment of ascites is presented in Table 5.1. Diuretics have now been used for treatment of fluid retention for more than 60 years. The diuretic treatment of ascites should start with an aldosterone antagonist [88]. These drugs act mainly on the distal tubules, where sodium-potassium transport takes place. If furosemide is administered alone, its effects on the loop of Henle may be counteracted by sodium reabsorbed in the distal tubules, because of secondary activation of aldosterone. The initial prescription in mild to moderate ascites should therefore be spironolactone 100 mg/day, a dose gradually increased up to 400 mg/day during the subsequent weeks [7, 88]. In order to avoid hyperkalaemia, a loop diuretic should be added before the full aldosterone antagonist dose is reached.

The effect of diuretic treatment often starts after 3-5 days of treatment. Body weight should be determined daily and serum concentration of sodium, potassium, and creatinine monitored. A daily weight loss of no more than 0.5–1.0 kg is recommended in order to prevent hypovolaemia [7]. In presence of gross peripheral oedema, a higher weight loss may be accepted. When weight loss is in progress, a loop diuretic should be added, often furosemide, as this drug may induce adequate diuresis and natriuresis. The recommended starting dose of furosemide is 40 mg/day and it can be raised to 160 mg/day with a stepwise increase every 2-3 days [88, 212]. Additional diuresis can sometimes be achieved by giving another diuretic in supplement, such as amiloride and thiazide, but side effects to these drugs are common and must be observed carefully.

The main goal of the treatment is to keep the patient free of ascites. When ascites has resolved, the diuretic dose should be reduced to a minimum and, if possible, discontinued. Non-steroid anti-inflammatory drugs (NSAIDs) often enhance sodium retention and formation of ascites, and they are associated with the development of renal failure, hyponatremia, and resistance to diuretic treatment. For these reasons, NSAID should be avoided [88]. Drugs that can induce arterial hypotension and therefore interfere with renal function, such as ACE inhibitors, angiotensin II antagonists, and α1-adrenergic blockers, should only be used very cautiously and exclusively when weight loss, blood pressure, serum sodium, potassium, and creatinine, and the clinical status are consistently monitored and adequate clinical reaction secured.

Table 5.1: Strategy in the treatment of sodium and fluid retention and complications in cirrhosis

Clinical State	Pathophysiologic Phase	Treatment
Preastic cirrhosis	1#	No specific treatment Moderat salt restriction (80-120 mmol/day, 4–7 g NaCl/day)
Mild ascites	2	Salt restriction (40-60 mmol/day, 2–3 g NaCl/day) Stepwise spironolactone (100-400 mg/day)
Moderate-tense ascites	3	Salt restriction Spironolactone Stepwise furosemide (40-160 mg/day)
Refractory ascites	3	Large volume paracentesis with volume substitution and diuretics TIPS (Transjugular porto-systemic stent)
Hyponatriaemia	3 and 4	Serum sodium < 125 mmol/l: diuretics discontinued Serum sodium < 120 mmol/l: volume expansion Hyper/euvolaemia: modest water restriction Experimental vasopressin V2-receptor antagonists
Hepatorenal syndrome	4 and 5	Vasoconstrictors and albumin TIPS Liver transplantation

Phase 1: Preascitic cirrhosis without activation of neurohumoral systems and intact renal function
Phase 2: Sodium-water retention with mild to moderate ascites and intact renal function
Phase 3: Activated neurohumoral systems, moderate to tense ascites, with or without renal impairment
Phase 4: Type-2 hepatorenal syndrome with activated neurohumoral systems and impaired renal function
Phase 5: Type-1 hepatorenal syndrome with activated neurohumoral systems and progressive renal impairment

5.1.2 TREATMENT OF REFRACTORY ASCITES

Approximately 10% of patients with ascites become refractory to medical treatment, and urge other treatment modalities [88, 212]. *Refractory ascites* is defined as diuretic-resistant ascites, which cannot be mobilized, even by intensive diuretic therapy and low salt diet (weight loss less than 200 g/day over 4 days); see Table 5.1. *Diuretic-intractable ascites* is characterized by the presence of diuretic-induced complications, such as encephalopathy and hyponatraemia [88]. The response to diuretics and the low salt diet should be validated only in clinically stable patients without other complications, like bleeding and infection. Such conditions, as earlier mentioned, substantially affect the ability to excrete salt and water.

Ascites recurs after a shorter or longer interval in the majority of patients, particularly in those with post-sinusoidal portal hypertension. Continuation of small dose of diuretics is therefore essential [301]. *Therapeutic paracentesis* should be performed in a single step and always combined with infusion of a plasma volume expander. The intraabdominal, vena caval, right atrial, and pulmonary capillary wedge pressures all decrease after large volume paracentesis [14]. Cardiac output may increase after 2-3 h and the mean arterial blood pressure decreases typically by 8-10 mmHg [171, 302]. Large volume paracentesis without adequate volume substitution may induce *post-paracentesis circulatory dysfunction* (PPCD) in up to 75% of the patients [89, 303]. PPCD is characterized by pronounced activation of the RAAS and SNS, which indicates aggravation of central hypovolaemia [303]. PPCD is mainly caused by paracentesis-induced splanchnic arteriolar vasodilatation and brings about a further reduction in systemic vascular resistance and may also give an increase in portal pressure [90, 145]. Several randomized, controlled trials of albumin infusion versus synthetic plasma expanders, such as dextran, collagen-based colloids, and starch, have shown equal effectiveness of plasma expanders in the prevention of post-paracentesis-induced complications after small volume paracentesis [304-307]. After paracentesis of large volume ascites (more than 5 l), albumin (8 g/l removed ascitic fluid) seems to be more effective than synthetic plasma expanders in preventing the paracentesis-induced increase in RAAS and volume related complications [88, 308, 309]. Intravenous albumin may therefore also prevent the development of other complications caused by circulatory dysfunction, such as HRS and rapid recurrence of ascites, and may also improve survival [89, 90, 145, 303]. For these reasons alternative plasma expanders should be avoided after large volume paracentesis, and reserved for the prevention of PPCD after small volume paracentesis [88].

In patients with tense ascites, which may cause abdominal discomfort, hemodynamic and respiratory dysfunction, therapeutic paracentesis should be preferred owing to fewer complications and a shorter stay in hospital compared to intensive diuretic treatment [23]. Several studies have shown that administration of vasoconstrictors such as terlipressin, midodrine, and noradrenaline may also be effective, either alone or in combination with albumin [153, 310, 311]. In a recent study, the vasoconstrictor midodrine was as effective as albumin in preventing PPCD, and at a lower cost [312]. This therapy has, however, been questioned in a newer small study of 24 patients [313].

PPCD is an example of a complication caused by further vasodilatation and reduction in central effective blood volume, which can be prevented by specific volume support and vasoconstriction. The use of specific vasoconstrictors such as midodrine or terlipressin may potentially play a role in the future strategy of treatment.

In the case of recurrent ascites, insertion of a TIPS should be considered. In experienced centers the technical success rate is high, at about 95% (314). Control of ascites is achieved in 80-90% of the cases, with complete resolution in 75% and TIPS is now considered more effective than repeated large volume paracentesis for the control of ascites [264, 315, 316]. A major challenge with insertion of TIPS is its relatively high frequency of hepatic encephalopathy, and, although patients often report a better quality of life, no significant effect on survival has been demonstrated after TIPS insertion [317]. However, improved survival after TIPS for refractory ascites has been demonstrated in a metaanalysis based on individual patient data from three large randomized trials [316]. The use of polytetrafluoroethylene-covered stents may improve shunt patency and thereby contribute to enhanced overall efficacy of this procedure [318]. Risk factors and relative contraindications for TIPS insertion include elevated serum bilirubin above 85 μmol/l, INR more than 2, episodic or chronic encephalopathy greater than grade 2, bacterial infections and renal, cardiac, and respiratory failure [174]. Although TIPS is more effective in controlling ascites compared with large volume paracentesis, it is at a cost of a higher frequency of hepatic encephalopathy, and it may not improve transplant-free survival. Large volume paracentesis with plasma expander infusion should therefore still be the first-time treatment for refractory ascites. TIPS should be regarded as a second-choice treatment in patients with well-preserved liver function and recurrence of ascites [7, 88].

5.1.3 TREATMENT OF HYPONATREMIA

Hyponatremia is defined as a reduction in serum sodium concentration below 130 mmol/l [300]. Hyponatremia often develops in cirrhosis, preferentially because of immense neuropituitary release of vasopressin. The presence of plasma volume expansion leads to hypervolaemic or dilutional hyponatremia [300]. Vasopressin acts on V_2 receptors (G protein coupled with cyclic AMP as second messenger) in the collecting ducts and controls the vasopressin-induced reabsorption of water [300]. This effect is mediated through selective water channels, aquaporins (AQPs), where AQP2 is the most important [234, 239]. Activation of AQPs increases water permeability, but in patients with ascites there is evidence of reduced AQP2 [222, 233, 239]. Clinical use of V_2-receptor antagonists, known as vaptans, may prove to be effective in the treatment of dilutional hyponatraemia [235, 249, 299]. Diuretics should be discontinued if serum sodium becomes less than 120 mmol/l.

5.2 TREATMENT AND PROPHYLAXIS OF SPONTANEOUS BACTERIAL PERITONITIS (SBP)

Culture from ascitic fluid often discloses bacterial species such as *Escherichia coli* and *Streptococcus* [293, 319]. Patients with cirrhosis and a low concentration of protein in the ascitic fluid (< 15 g/l) have an increased risk of development of SBP [320]. Antibiotic treatment of SBP significantly improves survival and third-generation cephalosporins, such as cefotaxime 2 g twice a day for up to 2 weeks, are often effective. Alternatively, amoxicillin/clavolanic acid can be considered. Fluoroquinolones are also effective. Other recommended antibiotics are ceftizoxime, cefonicid, ceftriaxone, and ceftazidime in adequate doses [88]. In patients with ascites and a history of SBP, prophylactic treatment with, for example, ciprofloxacin 250 mg/day orally, should be given, especially in patients with ascites with a low protein concentration [88, 122, 212, 293, 296]. Prophylactic treatment of patients with SBP with antibiotics alone, or better in combination with albumin, has significantly reduced the risk of development of type-1 HRS [297, 321]. Aminoglycosides should not be used, because of their toxic side effects.. Treatment with quinolones reduces the risk of development of SBP, but the effect on mortality is only marginal [322]. Patients without previous SBP but with ascitic protein < 15 g/l should be treated prophylactically with a quinolone such as ciprofloxacin 500 mg daily as long as ascites is present [88].

A new diagnostic paracentesis with cell count and cultures should be performed in patients with SBP after termination of antibiotic therapy. 70% of the patients, who recover after SBP, develop recurrence within the first year. Prophylactic treatment with quinolones after the first episode of SBP reduces the risk of a new SBP episode from 68% to 20% without clear effect on mortality [323, 324[. Several randomized studies have shown a decrease in hospitalizations and fewer recurrences of SBP in patients treated prophylactically. Patients with previous SBP should therefore be offered long-term prophylaxis with ciprofloxacin 500 mg daily. Alternatively, weekly treatment with 500-1000 mg ciprofloxacin could be considered [88, 296, 322].

5.3 TREATMENT OF HEPATORENAL SYNDROME (HRS)

Major elements in the development of HRS include hepatic insufficiency and a systemic circulatory dysfunction with a preferential renal vasoconstriction [8, 325, 326]. Hypothetically, the ideal drug would be a substance that improves liver function, reduces portal pressure, and exerts systemic splanchnic vasoconstriction, arterial volume expansion, and renal vasodilatation. Such a drug will probably never be developed, but these specific pathogenic mechanisms are each important targets for combined treatment. Possible renal and systemic target areas for pharmacologic intervention are summarized in Figure 4.3.

5.3.1 IMPROVEMENT OF LIVER DYSFUNCTION AND PORTAL HYPERTENSION

Liver transplantation is the ultimate treatment for HRS. Perioperatively, there may be a further deterioration of renal function, but within 1-2 months, GFR and RBF increase and haemodynamics and neurohumoral changes normalize [327, 328], and most patients with pretransplant kidney dysfunction do not experience progression to advanced dysfunction after liver transplantation [329]. In the waiting period for a liver transplantation, TIPS insertion has been used for portal decompression and central effective volume expansion in patients with HRS [88, 330].

5.3.2 CORRECTION OF CIRCULATORY DYSFUNCTION

The systemic administration of vasoconstrictors and plasma expanders in combination has shown beneficial effects on arterial vasodilatation, central effective hypovolemia, and renal dysfunction in patients with HRS [331, 332]. Terlipressin, a long-acting vasopressin analogue that stimulates splanchnic vasopressin V_{1a}-receptors, has been shown to increase arterial blood pressure, GFR, and urine volume in patients with HRS in a considerable number of patients [332-334]. Different studies have shown that terlipressin and albumin increase arterial blood pressure, suppress vasoconstrictor systems, and improve renal function in patients with HRS [332, 335]. Despite the dramatic haemodynamic effects of terlipressin, central and arterial blood volume increase only slightly after terlipressin administration, and the effect on central hypovolemia is only modest [326, 333]. In smaller studies, the combination of intravenous albumin and vasoconstrictors such as ornipressin, noradrenaline, dopamine, somatostatin, and octreotide have been shown to increase GRF and RBF, and normalize RAAS and SNS activity, although their effects are less potent than those of terlipressin [219, 336-338]. However, when combined with the α-adrenergic agonist midodrine, octreotide may have a short-term effect on RBF, GFR, and sodium excretion in a few patients with HRS [326, 339, 340]. In a study of 14 patients with type-1 HRS, the combination of midodrine, octreotide, and albumin significantly improved renal function [326, 341]. In a subset of the patients, TIPS insertion further improves renal function and sodium excretion for up to 12 months [341].

Recently, Ruiz-del-Arbol et al [135] demonstrated that decompensated patients with SBP and renal failure had lower cardiac output than those without renal failure, and the cardiac output further decreased in spite of antibiotic treatment. In these patients, renal failure might be precipitated as a mixed cirrhotic and septic cardiomyopathy [182, 342]. The HRS may thus develop as a combination of arterial vasodilatation, central effective hypovolemia, cardiac dysfunction, and renal vasoconstriction with renal hypoperfusion. Paracentesis and intravenous albumin should be considered in decompensated patients because it may improve renal perfusion by reducing the renal venous pressure [343]. However, a postparacentesis circulatory failure would have a negative effect on renal perfusion pressure because of a reduced arterial blood pressure, so a simultaneous infusion of albumin is important in these patients [145, 302, 344]. Treatment should be directed to support cardiac function and to treat any bacterial infections.

5.3.3 SUPPORT OF NEUROHUMORAL REGULATION

Central effective hypovolemia and arterial hypotension lead to a volume- and baroreceptor-induced activation of RAAS and the increased plasma renin activity correlates inversely with RBF and GFR [345, 346]. Infusion of pressor doses of angiotensin-II to decompensated patients increases renal perfusion and normalize arterial blood pressure in some patients, but it may have no or even harmful effects in others [23, 207]. Angiotensin-II mainly acts on the efferent arterioles, whereas afferent vasoconstriction is predominant in patients with HRS [346]. Low doses of the ACE inhibitor captopril induce a further reduction of GFR and filtration fraction, as well as of sodium excretion [88, 347]. Infusion of the angiotensin-II receptor antagonist losartan decreased portal pressure in patients with cirrhosis, but without significant effects on renal function [348-350]. In patients with HRS, the function of RAAS is essential in counteracting arterial hypotension, and the administration of inhibitors of the RAAS may have severe hypotensive action and further deteriorate renal and circulatory function [172].

Arginine-vasopressin is increased in cirrhosis and HRS primarily because of nonosmotic neuropituitary release [172]. Arginine-vasopressin induces vasoconstriction through V_1-receptors and renal tubular water retention through V2-receptors in the collecting ducts [351, 352]. In the kidneys, arginine-vasopressin acts on AQP2 in the collecting ducts and along with the secondary hyperaldosteronism contributes to the pronounced water reabsorption in patients with advanced HRS [222, 234, 239]. The action of arginine-vasopressin on renal vessels is limited, but systemic inhibition of V_1-receptors in experimental cirrhotic rats causes pronounced arterial hypotension [353, 354]. Administration of the vasopressin V2-receptor antagonist VPA/985 has been found to improve dilutional hyponatremia in patients with refractory ascites [355], but this treatment is still experimental [299].

5.4 TREATMENT OF BUDD-CHIARI SYNDROME AND VENO-OCCLUSIVE DISEASE

Budd-Chiari syndrome and veno-occlusive disease are both characterised by a posthepatic portal and sinusoidal hypertension. In contrast to cirrhosis, the sinusoidal lining is, at least in early disease, normal with a normal sieve-plate area of 10–20% [356, 357]. This means, that even a modest increase in the sinusoidal hydrostatic pressure gives a massive trans-sinusoidal filtration with formation of ascites with a high protein content. Painful ascites, suddenly developed, is a feauture that may be characteristic of both clinical syndromes. In some cases, there is a spontaneous or therapy-induced amelioration of the posthepatic occlusion, and ascites will disappear without further treatment. However, in several cases, following a more acute phase, ascites remains and the syndrome may end in hepatic cirrhosis or hepatic fibrosis [358, 359].

Diuretic treatment (see under Cirrhosis) in combination with TIPS-insertion and therapeutic paracentesis are the cornerstones of ascites treatment here. These patients seldomly develop hepatic encephalopathy and tolerate as a whole TIPS well [360]. Critically, the downstream loop of the TIPS-stent may be compromised by occlusive masses. By paracenthesis, it should be remembered that a substantial albumin and protein mass is removed as the concentration of these substances in the ascitic fluid is high, and an adequate intravenous supply of albumin should be secured.

5.5 PERSPECTIVE AND CONCLUSIONS ON TREATMENT OF HEPATIC ASCITES

Ascites and its complications, including SBP and HRS, are associated with poor prognoses despite treatment. However, our knowledge of the pathophysiology behind these severe complications has improved considerably, and there is now some optimism with respect to novel medical treatments in combination with different pharmacological approaches. The future may deal with different aspects in the pathophysiological process. A multitarget strategy should seek efficiently to counteract the arterial vasodilatation, central effective hypovolemia, and arterial hypotension by the administration of potent vasoconstrictors such as terlipressin in combination with plasma expanders like albumin. Development of long-acting systemic vasoconstrictors should be encouraged. Aquaporins may have a potential, especially in the treatment of dilutional hyponatremia. TIPS and β-blockers should be used to reduce portal pressure, whereas nitrates, COX-inhibitors, and enphrotoxic antibiotics should be used very cautiously. Cardiac function should be supported, especially in the presence of simultaneous infections.

5.6 TREATMENT OF NON-HEPATIC ASCITES

5.6.1 MALIGNENT ASCITES

Therapy should be directed against the primary tumor (e.g., ovarian cancer, GI-cancer, hepatoma, mesothelioma, diffuse carcinosis of the peritoneal lining). Surgical treatment, chemotherapy, radiation therapy, and the use of alpha- and beta-emitting isotopes locally are relevant but it is beyond the scope of this survey to go into details with such therapy. It should be stressed that decompressing manoeuvres like TIPS have no effect as the portal/sinusoidal pressure seldomly is increased. A small diuretic dose may sometimes (surprisingly) have effect in a few patients.

Paracenthesis with substitution of the high albumin/protein content in malignant ascites is indicated in order to relieve abdominal pain and respiratory discomfort and distress, and as a regular therapy in prolonged cases.

Reports on successful use of a peritoneal-venous shunt implant should be remembered, and this type of therapy is still considered in some centres.
The primary malignant disease determines the outcome, but with more prolonged and even chronic malignant cases, like in some gynaecological tumors and peritoneal mesothelioma, a more stationary approach must be considered.

5.6.2 CARDIAC ASCITES

In earlier times this was very common, but today cardiac ascites is most often seen in acute exacerbation of cardiac insufficiency and in the terminal phase of end-stage right heart congestive failure. Therapy is that of cardiac failure, and TIPS and peritoneal-venous shunting are of course contraindicated as the downstream drained areas have a higher hydrostatic pressure than the ascitic fluid (361).

Constrictive pericarditis should be treated by surgery but it should be remembered that a substantial diastolic dysfunction/insufficiency regularly persist. Cardiac transplantation is the ultimate cure, but is seldom possible.

5.6.3 INFECTIOUS ASCITES

Diagnostics should include search for abscesses, leaks, fistulas, local inflammation, etc., and intense therapy of these conditions must be performed whenever relevant. Bacterial diagnosis is described under subacute bacterial peritonitis (SBP), and in case of infectious ascites the white cell count is often very high [293]. Antibiotic therapy should be directed against the specific microorganisms, and in this context, it should be remembered that tuberculous ascites often present dramatic with a need of immediate treatment before results of TB-culture. In this case a laparoscopic biopsy with subsequent pathoanatomical visualisation of granulomas or acid-resistant bacteria, may give a clue to the diagnosis [88].

5.6.4 OTHER TYPES OF ASCITES

Hypoalbuminemic ascites (as in nephrotic syndrome, malnutrition, Kwashiorkor, beriberi), pancreatogenic ascites, and nephrogenic ascites should be treated specifically directed against the aetiology of the disease. Vigorous treatment of ascites is seldomly indicated, and in the few cases where ascites causes discomfort, symptomatic treatment should be directed against this.

5.7 CONCLUSION

The last decade has seen a substantial increase in the understanding of genesis and perpetuation of ascites and therapy based on pathophysiology. The majority of research has been directed towards pathophysiology and therapy of hepatic ascites. The complex pathophysiology of the substantially altered local mechanism in the dynamics of fluid exchange, and whole-body mechanisms with altered fluid distribution, systemic circulatory changes, major alterations in neurohumoral regulatory systems and cardiac and renal dysfunction of functional type in addition to liver disease, have disclosed the complex mechanisms involved in hepatic ascites and therapy of this syndrome. In contrast to ascites of other aetiologies large clinical controlled trials have substantially improved and evidence based the treatment of serious conditions, like the hepatorenal syndrome (HRS) and spontaneous bacterial peritonitis (SBP). There has been focus on correction of abnormal volume distribution and vasodilatation. Major questions like the exact mechanism of systemic vasodilatation, treatment of hyponatremia and dilutional fluid retention and cirrhotic cardiomyopathy still belongs to the future. Twenty-five years ago systemic effective hypovolemia with reduced central and arterial blood volume, arterial vasodilatation, sympathetic overactivity, and autonomic dysfunction was barely recognized and questioned by several. Systematic research performed at centers in for example Miami, New Haven, Denver, Calgary, London, Munich, Barcelona, Florence, and Copenhagen have by major experimental and clinical research contributed to our present knowledge. Several disciplines and research modalities have been involved, from basic animal experiments over clinical investigation to large clinical controlled trials of the multicentre type. These centres have worked locally as well as in an international collaboration that has been very fruitful to the benefit of patients with otherwise a very bad prognosis. Unlike much research in, for instance, clinical cardiology, hypertension, and diabetes, research in treatment and understanding of hepatic ascites has been investigator initiated, investigator driven and with a rather modest funding. Therefore, it is very pleasant to see the dramatic results of resent years research. Much work is still ahead, and should include cellular biology, signal transduction, membrane biochemistry, and other future disciplines in progress.

The general progress in understanding and treatment of heart insufficiency, malignant diseases and infectious diseases will result in progress in the understanding and treatment of non-hepatic ascites. Some types of non-hepatic ascites is owing to malnutrition, avitaminosis, low protein intake where pathophysiology and effective therapy is not a challenge due to lack of understanding, but due to lack of economy, development, organization, etc. It is the hope that the future will bring progress in both knowledge and means, so these very troublesome and dangerous clinical conditions, which have been known since ancient times, may be controlled and reduced to the benefit of numerous patients throughout the world.

References

[1] Michel CC, Phillips ME. Steady-state fluid filtration at different capillary pressures in perfused frog mesenteric capillaries. *J Physiol* 1987;388:421-435.

[2] Wiig H, Rubin K, Reed RK. New and active role of the interstitium in control of interstitial fluid pressure: potential therapeutic consequences. *Acta Anaesthesiol Scand* 2003;47(2):111-121. DOI: 10.1034/j.1399-6576.2003.00050.x

[3] Schrier RW. Water and sodium retention in edematous disorders: role of vasopressin and aldosterone. *Am J Med* 2006;119(7 Suppl 1):S47-S53. DOI: 10.1016/j.amjmed.2006.05.007

[4] Henriksen JH, Møller S, Ring-Larsen H, Christensen NJ. The sympathetic nervous system in liver disease. *J Hepatol* 1998;29(2):328-341. DOI: 10.1016/S0168-8278(98)80022-6

[5] Henriksen JH, Møller S. Alterations of hepatic and splanchnic microvascular exchange in cirrhosis: Local factors in the formation of ascites. In: Gines P, Arroyo V, Rodes J, Schrier RW, eds. *Ascites and renal dysfunction in liver disease.* 2nd ed. Malden: Blackwell; 2005. p. 174-185. DOI: 10.1002/9780470987476.ch14

[6] Gentilini P, Vizzutti F, Gentilini A, Zipoli M, Foschi M, Romanelli RG. Update on ascites and hepatorenal syndrome. *Dig Liver Dis* 2002;34(8):592-605. DOI: 10.1016/S1590-8658(02)80094-9

[7] Gines P, Cardenas A. The management of ascites and hyponatremia in cirrhosis. *Semin Liver Dis* 2008;28(1):43-58. DOI: 10.1055/s-2008-1040320

[8] Møller S, Henriksen JH, Bendtsen F. Pathogenetic background for treatment of ascites and hepatorenal syndrome. *Hepatol Int* 2008;2:416-428. DOI: 10.1007/s12072-008-9100-3

[9] Henriksen JH, Møller S, Schifter S, Abrahamsen J, Becker U. High arterial compliance in cirrhosis is related to elevated circulating calcitonin gene-related peptide (CGRP) and low adrenaline, but not to activated vasoconstrctor systems. *Gut* 2001;49:112-118. DOI: 10.1136/gut.49.1.112

[10] Henriksen JH, Kiszka-Kanowitz M, Bendtsen F, Møller S. Review article: volume expansion in patients with cirrhosis. *Aliment Pharmacol Ther* 2002;16 Suppl 5:12-23.

[11] Brinch K, Møller S, Bendtsen F, Becker U, Henriksen JH. Plasma volume expansion by albumin in cirrhosis. Relation to blood volume distribution, arterial compliance and severity of disease. *J Hepatol* 2003;39(1):24-31. DOI: 10.1016/S0168-8278(03)00160-0

[12] Iwakiri Y, Groszmann RJ. The hyperdynamic circulation of chronic liver diseases: From the patient to the molecule. *Hepatology* 2006;43(2 Suppl 1):S121-S131. DOI: 10.1002/hep.20993

[13] Schrier RW. Decreased effective blood volume in edematous disorders: what does this mean? *J Am Soc Nephrol* 2007;18(7):2028-2031. DOI: 10.1681/ASN.2006111302

[14] Møller S, Henriksen JH. Cardiovascular complications of cirrhosis. *Gut* 2008;57(2):268-278. DOI: 10.1136/gut.2006.112177

[15] Møller S, Henriksen JH, Bendtsen F. Ascites: Pathogenesis and therapeutic principles. *Scand J Gastroenterol* 2009;44:902-911.

[16] Barrowman JA, Granger DN. Effects of experimental cirrhosis on splanchnic microvascular fluid and solute exchange in the rat. *Gastroenterology* 1984;87(1):165-172.

[17] Joh T, Granger N, Bonoit JN. Intestinal microvascular responsiveness to norepinephrine in chronic portal hypertension. *Am J Physiol* 1991;260:H1135-H1143.

[18] Henriksen JH, Siemssen O, Krintel JJ, Malchow-Møller A, Bendtsen F, Ring-Larsen H. Dynamics of albumin in plasma and ascitic fluid in patients with cirrhosis. *J Hepatol* 2001;34(1):53-60. DOI: 10.1016/S0168-8278(00)00009-X

[19] Joffe P, Henriksen JH. Bidirectional peritoneal transport of albumin in continuous ambulatory peritoneal dialysis. *Nephrol Dial Transplant* 1995;10(9):1725-1732.

[20] Rippe B. Free water transport, small pore transport and the osmotic pressure gradient three-pore model of peritoneal transport. *Nephrol Dial Transplant* 2008;23(7):2147-2153. DOI: 10.1093/ndt/gfn049

[21] Lieberman FL, Denison EK, Reynolds TB. The relationship of plasma volume, portal hypertension, ascites, and renal sodium retention in cirrhosis. The overflow theory of ascites formation. *Ann NY Acad Sci* 1970;170:202-212. DOI: 10.1111/j.1749-6632.1970.tb37014.x

[22] Dagher L, Moore K. The hepatorenal syndrome. *Gut* 2001;49(5):729-737. DOI: 10.1136/gut.49.5.729

[23] Arroyo V, Colmenero J. Ascites and hepatorenal syndrome in cirrhosis: pathophysiological basis of therapy and current management. *J Hepatol* 2003;38 Suppl 1:S69-S89. DOI: 10.1016/S0168-8278(03)00007-2

[24] Gines P, Cardenas A, Arroyo V, Rodes J. Management of cirrhosis and ascites. *N Engl J Med* 2004;350(16):1646-1654. DOI: 10.1056/NEJMra035021

[25] Henriksen JH, Winkler K. Peritoneum and ascites formation. In: Bengsmark S, ed. *The peritoneum and peritoneal access.*London: Wright; 1989. p. 94-110.

[26] Dawson AD. Historical notes on ascites. *Gastroenterology* 1960;39:790-791.

[27] Jarcho S. Ascites as described by Aulus Cornelius Celsus (ca. A. D. 30). *Am J Cardiol* 1958;2(4):507-508. DOI: 10.1016/0002-9149(58)90339-4

[28] Henriksen JH. Ernest Henry Starling. *Physician and physiologis*t. 1 ed. Copenhagen: Lægeforeningens Forlag; 2000.

[29] Starling EH, Tubby AH. On the absorption from and secretion into the serous cavities. *J Physiol* 1894;16(140):155.

[30] Bayliss WM, Starling EH. Observations on venous pressures and their relationships to capillary pressure. *J Physiol* 1894;16:159-202.

[31] Starling EH. The influence of mechanical factors on lymph production. *J Physiol* 1894;16:224-267.

[32] Leathes JB, . On the absorption of salt solution from the plural cavities. *J Physiol* 1895;18:106-116.

[33] Starling EH. On the absorption of fluids from the connective tissure spaces. *J Physiol* 1896;19:312-326.

[34] Rubin JC, Novak J, Squire JJ. Ovarian fibromas and theca-cell tumors: report of 78 cases with special reference to ptoduction of ascites and hydrothorax (Meigs syndrome). *Am J Obst Gyn* 1944;48:601-616.

[35] Henriksen JH, Horn T, Christoffersen P. The blood-lymph barrier in the liver. A review based on morphological and functional concepts of normal and cirrhotic liver. *Liver* 1984; 4(4):221-232. DOI: 10.1111/j.1600-0676.1984.tb00932.x

[36] Henriksen JH. Colloid osmotic pressure in decompensated cirrhosis. A 'mirror image' of portal venous hypertension. *Scand J Gastroenterol* 1985;20(2):170-174. DOI: 10.3109/00365528509089651

[37] McHugh PP, Shah SH, Johnston TD, Gedaly R, Ranjan D. Predicting dry weight in patients with ascites and liver cirrhosis using computed tomography imaging. *Hepato-gastroenterology* 2010;57:591-597.

[38] Jeong SH, Lee JA, Kim JA, Lee MW, Chae HB, Choi WJ, Shin HS, Lee KH, Youn SJ, Koong SS, Park SM. Assessment of body composition using dual energy x-ray absorpti-

ometry in patients with liver cirrhosis: comparison with anthropometry. *Korean J Intern Med* 1999;14(2):64-71.

[39] Henriksen JH. Variability of hydrostatic hepatic vein and ascitic fluid pressure, and of plasma and ascitic fluid colloid osmotic pressure in patients with liver cirrhosis. *Scand J Clin Lab Invest* 1980;40(6):515-522. DOI: 10.3109/00365518009091958

[40] Henriksen JH, Schlichting P. Increased extravasation and lymphatic return rate of albumin during diuretic treatment of ascites in patients with liver cirrhosis. Scand J Clin Lab Invest 1981;41(6):589-599. DOI: 10.3109/00365518109090503

[41] Michel CC. Transport of macromolecules through microvascular walls. *Cardiovasc Res* 1996;32(4):644-653.

[42] Wright DM, Wiig H, Winlove CP, Bert JL, Reed RK. Simultaneous measurement of interstitial fluid pressure and load in rat skin after strain application in vitro. *Ann Biomed Eng* 2003;31(10):1246-1254. DOI: 10.1114/1.1616933

[43] Levick JR. Revision of the Starling principle: new views of tissue fluid balance. *J Physiol* 2004;557:704-711. DOI: 10.1113/jphysiol.2004.066118

[44] Laine GA, Hall JT, Laine SH, Granger J. Transsinusoidal fluid dynamics in canine liver during venous hypertension. *Circ Res* 1979;45(3):317-323. DOI: 10.1161/01.RES.45.3.317

[45] Witte CL, Witte MH, Dumont AE. Lymph imbalance in the genesis and perpetuation of the ascites syndrome in hepatic cirrhosis. *Gastroenterology* 1980;78:1059-1068.

[46] Leak LV. Lymphatic endothelial-interstitial interface. *Lymphology* 1987;20(4):196-204.

[47] Huet PM, Goresky CA, Villeneuve JP, Marleau D, Lough JO. Assessment of liver microcirculation in human cirrhosis. *J Clin Invest* 1982;70(6):1234-1244. DOI: 10.1172/JCI110722

[48] Goresky CA. The 1994 G. Malcolm Brown Lecture. Biological barriers and medicine. *Clin Invest Med* 1995;18(6):484-501.

[49] Villeneuve JP, Dagenais M, Huet PM, Roy A, Lapointe R, Marleau D. The hepatic microcirculation in the isolated perfused human liver. *Hepatology* 1996;23:24-31. DOI: 10.1002/hep.510230104

[50] Braet F, Wisse E. Structural and functional aspects of liver sinusoidal endothelial cell fenestrae: a review. *Comp Hepatol* 2002;1(1):1. DOI: 10.1186/1476-5926-1-1

[51] Goresky CA. Kinetic interpretation of hepatic multiple-indicator dilution studies. *Am J Physiol* 1983;245(1):G1-12.

[52] Goresky CA, Simard A, Schwab AJ. Increased hepatocyte permeability surface area product for 86Rb with increase in blood flow. *Circ Res* 1997;80(5):645-654. DOI: 10.1161/01.RES.80.5.645

[53] Goresky CA. The modeling of tracer exchange and sequestration in the liver. *Fed Proc* 1984;43(2):154-160.

[54] Larsen OA, Winkler K, Tygstrup N. Extra plasma in the liver calculated from the hepatic hematocrit in patients with portacaval anastomosis. *Clin Sci* 1963;25:357-360.

[55] Oikawa H, Masuda T, Kawaguchi J, Sato R. Three-dimensional examination of hepatic stellate cells in rat liver and response to endothelin-1 using confocal laser scanning microscopy. *J Gastroenterol Hepato*l 2002;17(8):861-872. DOI: 10.1046/j.1440-1746.2002.02831.x

[56] Rippe B, Rosengren BI, Venturoli D. The peritoneal microcirculation in peritoneal dialysis. *Microcirculation* 2001;8(5):303-320.

[57] Dichi JB, Dichi I, Maio R, Correa CR, Angeleli AY, Bicudo MH, Rezende TA, Burini RC. Whole-body protein turnover in malnourished patients with child class B and C cirrhosis on diets low to high in protein energy. *Nutrition* 2001;17(3):239-242. DOI: 10.1016/S0899-9007(00)00567-0

[58] Krediet RT, Struijk DG. Peritoneal dialysis membrane evaluation in clinical practice. *Contrib Nephrol* 2012;178:232-237. DOI: 10.1159/000337884

[59] Krediet RT. Peritoneal physiology--impact on solute and fluid clearance. *Adv Ren Replace Ther* 2000;7(4):271-279. DOI: 10.1053/jarr.2000.16269

[60] Levick JR, Michel CC. Microvascular fluid exchange and the revised Starling principle. *Cardiovasc Res* 2010;87(2):198-210. DOI: 10.1093/cvr/cvq062

[61] Richardson PD, Granger DN, Taylor AE. Capillary filtration coefficient: the technique and its application to the small intestine. *Cardiovasc Res* 1979;13(10):547-561. DOI: 10.1093/cvr/13.10.547

[62] Korthuis RJ, Benoit JN, Kvietys PR, Townsley MI, Taylor AE, Granger N. Humoral factors may mediate increased rat hindquater blood flow in portal hypertension. *Am J Physiol* 1985;249:H827-H833.

[63] Schrier RW, Arroyo V, Bernardi M, Epstein M, Henriksen JH, Rodés J. Peripheral artery vasodilatation hypothesis: A proposal for the initiation of renal sodium and water retention in cirrhosis. *Hepatology* 1988;5:1151-1157. DOI: 10.1002/hep.1840080532

[64] Ekataksin W, Kaneda K. Liver microvascular architecture: an insight into the pathophysiology of portal hypertension. *Semin Liver Dis* 1999;19(4):359-382. DOI: 10.1055/s-2007-1007126

[65] Groszmann RJ. Vasodilatation and hyperdynamic circulatory state in chronic liver disease. In: Bosch J, Groszmann RJ, eds. *Portal hypertension. Pathophysiology and treatment.* 1 ed. Oxford: Blackwell; 1994. p. 17-26.

[66] Colombato LA, Albillos A, Groszmann RJ. Role of central blood volume in the development of sodium retention in portal hypertensive rats. *Hepatology* 1993;18:100A. DOI: 10.1016/0270-9139(93)91930-Q

[67] Lebrec D, Fleury PD, Rueff B, Nahum H, Benhamou JP. Portal hypertension, size of esophageal varices, and risk of gastrointestinal bleeding in alcoholic cirrhosis. *Gastroenterology* 1980;79:1139-1144.

[68] Popper H, Schaffner F. Fine structural changes of the liver. *Ann Intern Med* 1963;59:674-91.:674-691.

[69] Moragas A, Allende H, Sans M. Characteristics of perisinusoidal collagenization in liver cirrhosis: computer-assisted quantitative analysis. *Anal Quant Cytol Histol* 1998;20(3):169-177.

[70] Schaffner F, Popper H. Morphologic studies in neonatal cholestasis with emphasis on giant cells. *Ann N Y Acad Sci* 1963;111:358-374.

[71] Schaffner F. The history of liver disease at The Mount Sinai Hospital. *Mt Sinai J Med* 2000;67(1):76-83.

[72] Salo J, Gines A, Gines P, Piera C, Jimenez W, Guevara M, Fernandezesparrach G, Sort P, Bataller R, Arroyo V, Rodes J. Effect of therapeutic paracentesis on plasma volume and transvascular escape rate of albumin in patients with cirrhosis. *J Hepatol* 1997;27:645-653. DOI: 10.1016/S0168-8278(97)80081-5

[73] Witte MH, Witte CL, Dumont AE. Estimated net transcapillary water and protein flux in the liver and intestine of patients with portal hypertension from hepatic cirrhosis. *Gastroenterology* 1981;80(2):265-272.

[74] Lassen NA, Perl W. *Tracer kinetics methods in medical physiology.* New York: Raven; 1979.

[75] Rippe B, Perry MA, Granger DN. Permselectivity of the peritoneal membrane. *Microvasc Res* 1985;29(1):89-102. DOI: 10.1016/0026-2862(85)90009-3

[76] Benoit JN, Barrowman JA, Harper SL, Kvietys PR, Granger DN. Role of humoral factors in the intestinal hyperemia associated with chronic portal hypertension. *Am J Physiol* 1984;247:G486-G493.

[77] Benoit JN, Granger DN. Splanchnic hemodynamics in chronic portal hypertension. *Seminars in Liver disease* 1986;6(4):287-298. DOI: 10.1055/s-2008-1040611

[78] Runyon BA. Management of adult patients with ascites due to cirrhosis: An update. *Hepatology* 2009;49(6):2087-2107. DOI: 10.1002/hep.22853

[79] Joffe P, Henriksen JH. Aspects of osseous, peritoneal and renal handling of bisphosphonate during peritoneal dialysis: a methodological study. *Scand J Clin Lab Invest* 1996;56(4):327-337. DOI: 10.3109/00365519609090584

[80] Carlsson O, Nielsen S, Zakaria e, Rippe B. In vivo inhibition of transcellular water channels (aquaporin-1) during acute peritoneal dialysis in rats. *Am J Physiol* 1996;271(6 Pt 2):H2254-H2262.

[81] Schnitzer JE, McIntosh DP, Dvorak AM, Liu J, Oh P. Separation of caveolae from associated microdomains of GPI-anchored proteins. *Science* 1995;269(5229):1435-1439. DOI: 10.1126/science.7660128

[82] Akriviadis AA, Kapniasis D, Hadjigavriel M, Mitsiou A, Goulis J. Serum/ascites albumin gradient. Its value as a rational approach to the differential diagnosis of ascites. *Scand J Gastroenterol* 1996;31:814-817. DOI: 10.3109/00365529609010358

[83] Henriksen JH. Cirrhosis: Ascites and hepatorenal syndrome. Recent advances in pathogenesis. *J Hepatol* 1995;23:25-30.

[84] Granger DN, Granger JP, Brace RA, Parker RE, Taylor AE. Analysis of the permeability characteristics of cat intestinal capillaries. *Circ Res* 1979;44(3):335-344. DOI: 10.1161/01.RES.44.3.335

[85] Schrier RW, Gurevich AK, Cadnapaphornchai MA. Pathogenesis and management of sodium and water retention in cardiac failure and cirrhosis. *Semin Nephrol* 2001;21(2):157-172. DOI: 10.1053/snep.2001.20933

[86] Epstein M. Peritoneovenous shunt in the management of ascites and the hepatorenal syndrome. In: Epstein M, ed. *The Kidney in Liver Disease.* 4 ed. Philadelphia: Hanley and Belfus; 1996. p. 491-506.

[87] Gines A, Planas R, Angeli P, Guarner C, Salerno F, Gines P, Salo J, Rodriguez N, Domenech E, Soriano G, Anibarro L, Gassull MA, Gatta A, Arroyo V, Rodes J. Treatment

of patients with cirrhosis and refractory ascites using LeVeen shunt with titanium tip: Comparison with therapeutic paracentesis. *Hepatology* 1995;22:124-131.

[88] Gines P, Angeli P, Lenz K, Møller S, Moore K, Moreau R, Merkel C, Ring-Larsen H, Bernardi M, Garcia-Tsao G, Hayes P. EASL clinical practice guidelines on the management of ascites, spontaneous bacterial peritonitis, and hepatorenal syndrome in cirrhosis. *J Hepatol* 2010;53(3):397-417. DOI: 10.1016/j.jhep.2010.05.004

[89] Gines P, Guevara M, De Las HD, Arroyo V. Review article: albumin for circulatory support in patients with cirrhosis. *Aliment Pharmacol Ther* 2002;16:24-31. DOI: 10.1046/j.1365-2036.16.s5.4.x

[90] Bernardi M, Caraceni P, Navickis RJ, Wilkes MM. Albumin infusion in patients undergoing large-volume paracentesis: A meta-analysis of randomized trials. *Hepatology* 2012;55(4):1172-1181. DOI: 10.1002/hep.24786

[91] Møller S, Henriksen JH. Endothelins in chronic liver disease. *Scand J Clin Lab Invest* 1996;56:481-490. DOI: 10.3109/00365519609088803

[92] Tazi KA, Barriere E, Moreau R, Heller J, Sogni P, Pateron D, Poirel O, Lebrec D. Role of shear stress in aortic eNOS up-regulation in rats with biliary cirrhosis. *Gastroenterology* 2002;122(7):1869-1877. DOI: 10.1053/gast.2002.33586

[93] Wiest R, Groszmann RJ. The paradox of nitric oxide in cirrhosis and portal hypertension: too much, not enough. *Hepatology* 2002;35(2):478-491. DOI: 10.1053/jhep.2002.31432

[94] Zipprich A, Mehal WZ, Ripoll C, Groszmann RJ. A distinct nitric oxide and adenosine A1 receptor dependent hepatic artery vasodilatatory response in the CCl-cirrhotic liver. *Liver Int* 2010;7:988-994. DOI: 10.1111/j.1478-3231.2010.02278.x

[95] La Villa G, Barletta G, Pantaleo P, Del Bene R, Vizzutti F, Vecchiarino S, Masini E, Perfetto F, Tarquini R, Gentilini P, Laffi G. Hemodynamic, renal, and endocrine effects of acute inhibition of nitric oxide synthase in compensated cirrhosis. *Hepatology* 2001;34(1):19-27. DOI: 10.1053/jhep.2001.25756

[96] Wiest R, Shah V, Sessa WC, Groszmann RJ. NO overproduction by eNOS precedes hyperdynamic splanchnic circulation in portal hypertensive rats. *Am J Physiol* 1999;276(4 Pt 1):G1043-G1051.

[97] Cardenas A, Lowe R, Oh S, Bodkin S, Kenney T, Lamorte WW, Afdhal NH. Hemodynamic effects of substance P and its receptor antagonist RP67580 in anesthetized rats with carbon tetrachloride-induced cirrhosis. *Scand J Gastroenterol* 2008; 43(3):228-233. DOI: 10.1080/00365520701685691

[98] Møller S, Bendtsen F, Henriksen JH. Vasoactive substances in the circulatory dysfunction of cirrhosis. *Scand J Clin Lab Invest* 2001;61(6):421-429. DOI: 10.1080/00365510152567059

[99] Møller S, Gülberg V, Henriksen JH, Gerbes AL. Endothelin-1 and endothelin-3 in cirrhosis: relations to systemic and splanchnic haemodynamics. *J Hepatol* 1995;23:135-144. DOI: 10.1016/0168-8278(95)80327-0

[100] Møller S, Bendtsen F, Schifter S, Henriksen JH. Relation of calcitonin gene-related peptide to systemic vasodilatation and central hypovolaemia in cirrhosis. *Scand J Gastroenterol* 1996;31:928-933. DOI: 10.3109/00365529609052004

[101] Lewis FW, Cohen JA, Rector WG. Autonomic dysfunction in alcoholic cirrhosis: Relationship to indicators of sympathetic activation and the occurrence of renal sodium retention. *Am J Gastroenterol* 1991;86:553-559.

[102] Salerno F, Cazzaniga M. Autonomic dysfunction: often present but usually ignored in patients with liver disease. *Liver Int* 2009;29(10):1451-1453. DOI: 10.1111/j.1478-3231.2009.02141.x

[103] Møller S, Mortensen C, Bendtsen F, Jensen LT, Gotze JP, Madsen JL. Cardiac sympathetic imaging with mIBG in cirrhosis and portal hypertension: Relation to autonomic and cardiac function. *Am J Physiol Gastrointest Liver Physiol* 2012;303:G1228-G1235.

[104] Braillon A, Moreau R, Hadengue A, Roulot D, Sayegh R, Lebrec D. Hyperkinetic circulatory syndrome in patients with presinusoidal portal hypertension.Effect of propranolol. *J Hepatol* 1989;9:312-318. DOI: 10.1016/0168-8278(89)90139-6

[105] Albillos A, Banáres R, Barrios C, Clemente G, Rossi I, Escartin P, Bosch J. Oral administration of clonidine in patients with alcoholic cirrhosis. *Gastroenterology* 1992;102:248-254.

[106] Henriksen JH, Bendtsen F, Gerbes AL, Christensen NJ, Ring-Larsen H, Sørensen TIA. Estimated central blood volume in cirrhosis - Relationship to sympathetic nervous activity, beta-adrenergic blockade and atrial natriuretic factor. *Hepatology* 1992;16:1163-1170. DOI: 10.1002/hep.1840160510

[107] Reiberger T, Ulbrich G, Ferlitsch A, Payer BA, Schwabl P, Pinter M, Heinisch BB, Trauner M, Kramer L, Peck-Radosavljevic M. Carvedilol for primary prophylaxis of variceal bleeding in cirrhotic patients with haemodynamic non-response to propranolol. *Gut* 2013.

[108] Hobolth L, Møller S, Gronbaek H, Roelsgaard K, Bendtsen F, Feldager HE. Carvedilol or propranolol in portal hypertension? A randomized comparison. *Scand J Gastroenterol* 2012;47(4):467-474. DOI: 10.3109/00365521.2012.666673

[109] Pinzani M, Milani S, Defranco R, Grappone C, Caligiuri A, Gentilini A, Tostiguerra C, Maggi M, Failli P, Ruocco C, Gentilini P. Endothelin 1 is overexpressed in human cirrhotic liver and exerts multiple effects on activated hepatic stellate cells. *Gastroenterology* 1996;110:534-548. DOI: 10.1053/gast.1996.v110.pm8566602

[110] Pinzani M, Marra F, Carloni V. Signal transduction in hepatic stellate cells. *Liver* 1998;18(1):2-13. DOI: 10.1111/j.1600-0676.1998.tb00120.x

[111] Rockey DC. Hepatic blood flow regulation by stellate cells in normal and injured liver. *Semin Liver Dis* 2001;21(3):337-349. DOI: 10.1055/s-2001-17551

[112] Reynaert H, Thompson MG, Thomas T, Geerts A. Hepatic stellate cells: role in microcirculation and pathophysiology of portal hypertension. *Gut* 2002;50(4):571-581. DOI: 10.1136/gut.50.4.571

[113] Henriksen JH, Gulberg V, Fuglsang S, Schifter S, Bendtsen F, Gerbes AL, Møller S. Q-T interval (QT(C)) in patients with cirrhosis: relation to vasoactive peptides and heart rate. *Scand J Clin Lab Invest* 2007;67:643-653. DOI: 10.1080/00365510601182634

[114] Wiest R. Splanchnic and systemic vasodilation: the experimental models. *J Clin Gastroenterol* 2007;41(10 Suppl 3):S272-S287. DOI: 10.1097/MCG.0b013e318157cb57

[115] Iwakiri Y. Endothelial dysfunction in the regulation of cirrhosis and portal hypertension. *Liver Int* 2012; 32(2):199-213.

[116] Shah V. Cellular and molecular basis of portal hypertension. *Clin Liver Dis* 2001;5(3):629-644. DOI: 10.1016/S1089-3261(05)70185-9

[117] Reed RK, Wiig H. Interstitial albumin mass and transcapillary extravasation rate of albumin in DMBA-induced rat mammary tumors. *Scand J Clin Lab Invest* 1983;43 (6):503-512. DOI: 10.3109/00365518309168437

[118] de Vries EG, Beekhuis H, de Jong R, Mulder NH, Piers DA. Evidence for capillary leakage during chemotherapy in man. *Eur J Clin Invest* 1986;16(3):243-247. DOI: 10.1111/j.1365-2362.1986.tb01336.x

[119] Fleck A, Raines G, Hawker F, Trotter J, Wallace PI, Ledingham IM, Calman KC. Increased vascular permeability: a major cause of hypoalbuminaemia in disease and injury. *Lancet* 1985;1(8432):781-784. DOI: 10.1016/S0140-6736(85)91447-3

[120] Bronskill MJ, Bush RS, Ege GN. A quantitative measurement of peritoneal drainage in malignant ascites. *Cancer* 1977;40(5):2375-2380.

[121] Deysine M, Stein S. Albumin shifts across the extracellular space secondary to experimental infections. *Surg Gynecol Obstet* 1980;151(5):617-620.

[122] Bernardi M. Spontaneous bacterial peritonitis: from pathophysiology to prevention. *Intern Emerg Med* 2010;5 Suppl 1:S37-44.:S37-S44.

[123] Morgan AG, Terry SI. Impaired peritoneal fluid drainage in nephrogenic ascites. *Clin Nephrol* 1981;15(2):61-65.

[124] Ing TS, Daugirdas JT, Popli S, Kheirbek AO, Gandhi VC. Treatment of refractory hemodialysis ascites with maintenance peritoneal dialysis. *Int J Artif Organs* 1980;3 (5):311.

[125] Kavoliuniene A, Vaitiekiene A, Cesnaite G. Congestive hepatopathy and hypoxic hepatitis in heart failure: A cardiologist's point of view. *Int J Cardiol* 2013;166(3):554-558.

[126] Witte CL, Witte MH. The portocardiorenal axis and refractory ascites: The underfilled cup runneth over. *Hepatology* 1993;10(1):114-119. DOI: 10.1002/hep.1840100122

[127] Francis GS. Pathophysiology of chronic heart failure. *Am J Med* 2001;110 Suppl 7A: 37S-46S. DOI: 10.1016/S0002-9343(98)00385-4

[128] Schrier RW, Abraham WT. Hormones and hemodynamics in heart failure. *N Engl J Med* 1999;341(8):577-585. DOI: 10.1056/NEJM199908193410806

[129] Møller S, Dumcke CW, Krag A. The heart and the liver. *Expert Rev Gastroenterol Hepatol* 2009; 3(1):51-64. DOI: 10.1586/17474124.3.1.51

[130] Bongartz LG, Cramer MJ, Doevendans PA, Joles JA, Braam B. The severe cardiorenal syndrome: 'Guyton revisited'. *Eur Heart J* 2005;26(1):11-17. DOI: 10.1093/eurheartj/ehi020

[131] Guyton AC. *Textbook of medical physiology.* 8 ed. Philadelphia: W.B.Saunders; 1991.

[132] Langer DA, Shah VH. Nitric oxide and portal hypertension: Interface of vasoreactivity and angiogenesis. *J Hepatol* 2006;44(1):209-216. DOI: 10.1016/j.jhep.2005.10.004

[133] Schrier RW, Niederberger M, Weigert A, Gines P. Peripheral arterial vasodilatation:-Determinant of functional spectrum of cirrhosis. *Sem Liver Dis* 1994;14:14-22. DOI: 10.1055/s-2007-1007294

[134] Hadengue A, Moreau R, Gaudin C, Bacq Y, Champigneulle B, Lebrec D. Total effective vascular compliance in patients with cirrhosis: a study of the response to acute blood volume expansion. *Hepatology* 1992;15:809-815. DOI: 10.1002/hep.1840150511

[135] Ruiz-Del-Arbol L, Urman J, Fernandez J, Gonzalez M, Navasa M, Monescillo A, Albillos A, Jimenez W, Arroyo V. Systemic, renal, and hepatic hemodynamic derangement in cirrhotic patients with spontaneous bacterial peritonitis. *Hepatology* 2003;38(5):1210-1218. DOI: 10.1053/jhep.2003.50447

[136] Caraceni P, Dazzani F, Salizzoni E, Domenicali M, Zambruni A, Trevisani F, Bernardi M. Muscle circulation contributes to hyperdynamic circulatory syndrome in advanced cirrhosis. *J Hepatol* 2008;48:559-566. DOI: 10.1016/j.jhep.2007.12.016

[137] Henriksen JH, Ring-Larsen H, Kanstrup I-L, Christensen NJ. Splanchnic and renal elimination and release of catecholamines in cirrhosis. Evidence of enhanced sympathetic nervous activity in patients with cirrhosis. *Gut* 1984;25:1034-1043. DOI: 10.1136/gut.25.10.1034

[138] Møller S, Bendtsen F, Henriksen JH. Determinants of the renin-angiotensin-aldosterone system in cirrhosis with special emphasis on the central blood volume. *Scand J Gastroenterol* 2006;451-458. DOI: 10.1080/00365520500292962

[139] Møller S, Henriksen JH. Neurohumoral fluid regulation in chronic liver disease. *Scand J Clin Lab Invest* 1998;58(5):361-372. DOI: 10.1080/00365519850186346

[140] Epstein M. Renal sodium handling in liver disease. In: Epstein M, ed. *The Kidney in Liver Disease.* 4 ed. Philadelphia: Hanley and Belfus; 1996. p. 1-31.

[141] Gerbes AL, Møller S, Gülberg V, Henriksen JH. Endothelin-1 and -3 plasma concentrations in patients with cirrhosis: Role of splanchnic and renal passage and liver function. *Hepatology* 1995;21:735-739.

[142] Schrier RW, Shchekochikhin D, Gines P. Renal failure in cirrhosis: prerenal azotemia, hepatorenal syndrome and acute tubular necrosis. *Nephrol Dial Transplant* 2012;27(7):2625-2628. DOI: 10.1093/ndt/gfs067

[143] Henriksen JH, Ring-Larsen H. Raised renal venous pressure: Direct cause of renal sodium retention in cirrhosis? *Lancet* 1988;ii:8602.

[144] Møller S, Bendtsen F, Henriksen JH. Effect of volume expansion on systemic hemodynamics and central and arterial blood volume in cirrhosis. *Gastroenterology* 1995;109:1917-1925. DOI: 10.1016/0016-5085(95)90759-9

[145] Sola-Vera J, Minana J, Ricart E, Planella M, Gonzalez B, Torras X, Rodriguez J, Such J, Pascual S, Soriano G, Perez-Mateo M, Guarner C. Randomized trial comparing albumin and saline in the prevention of paracentesis-induced circulatory dysfunction in cirrhotic patients with ascites. *Hepatology* 2003;37(5):1147-1153. DOI: 10.1053/jhep.2003.50169

[146] Kiszka-Kanowitz M, Henriksen JH, Møller S, Bendtsen F. Blood volume distribution in patients with cirrhosis: aspects of the dual-head gamma-camera technique. *J Hepatol* 2001;35(5):605-612. DOI: 10.1016/S0168-8278(01)00175-1

[147] Henriksen JH, Fuglsang S, Bendtsen F, Møller S. Arterial hypertension in cirrhosis: arterial compliance, volume distribution, and central haemodynamics. *Gut* 2006;380-387. DOI: 10.1136/gut.2005.064329

[148] Andreu V, Perello A, Moitinho E, Escorsell A, Garcia-Pagan JC, Bosch J, Rodes J. Total effective vascular compliance in patients with cirrhosis. Effects of propranolol. *J Hepatol* 2002;36(3):356-361. DOI: 10.1016/S0168-8278(01)00300-2

[149] Henriksen JH, Bendtsen F, Sørensen TIA, Stadeager C, Ring-Larsen H. Reduced central blood volume in cirrhosis. *Gastroenterology* 1989;97:1506-1513.

[150] Møller S, Søndergaard L, Møgelvang J, Henriksen O, Henriksen JH. Decreased right heart blood volume determined by magnetic resonance imaging: Evidence of central underfilling in cirrhosis. *Hepatology* 1995;22:472-478.

[151] Møller S, Henriksen JH, Bendtsen F. Central- and non-central blood volumes in cirrhosis.Relation to anthropometrics and gender. *Am J Physiol Gastrointest Liver Physiol* 2003;284:G970-G979.

[152] Schrier RW. Renin-angiotensin in preascitic cirrhosis: Evidence for primary peripheral arterial vasodilatation. *Gastroenterology* 1998;115(2):489-491. DOI: 10.1016/S0016-5085(98)70215-X

[153] Krag A, Møller S, Henriksen JH, Holstein-Rathlou NH, Larsen FS, Bendtsen F. Terlipressin improves renal function in patients with cirrhosis and ascites without hepatorenal syndrome. *Hepatology* 2007;46(6):1863-1871. DOI: 10.1002/hep.21901

[154] Møller S, Nørgaard A, Henriksen JH, Frandsen E, Bendtsen F. Effects of tilting on central hemodynamics and homeostatic mechanisms in cirrhosis. *Hepatology* 2004;40(4):811-819.

[155] Ruiz-Del-Arbol L, Monescillo A, Jimenez W, Garcia-Plaza A, Arroyo V, Rodes J. Paracentesis-induced circulatory dysfunction: mechanism and effect on hepatic hemodynamics in cirrhosis. *Gastroenterology* 1997;113(2):579-586. DOI: 10.1053/gast.1997.v113.pm9247479

[156] Henriksen JH, Schütten HJ, Bendtsen F, Warberg J. Circulating atrial natriuretic peptide (ANP) and central blood volume (CBV) in cirrhosis. *Liver* 1986;6:361-368. DOI: 10.1111/j.1600-0676.1986.tb00305.x

[157] Henriksen JH, Gulberg V, Gerbes AL, Bendtsen F, Møller S. Increased arterial compliance in cirrhosis is related to decreased arterial C-type natriuretic peptide, but not to atrial natriuretic peptide. *Scand J Gastroenterol* 2003;38(5):559-564. DOI: 10.1080/00365520310000393

[158] Møller S, Henriksen JH. Cirrhotic cardiomyopathy. *J Hepatol* 2010;53(1):179-190. DOI: 10.1016/j.jhep.2010.02.023

[159] Arroyo V, Terra C, Gines P. Advances in the pathogenesis and treatment of type-1 and type-2 hepatorenal syndrome. *J Hepatol* 2007;46(5):935-946. DOI: 10.1016/j.jhep.2007.02.001

[160] Albillos A, Colombato LA, Groszmann RJ. Vasodilatation and sodium retention in prehepatic portal hypertension. *Gastroenterology* 1992;102:931-935.

[161] Colombato LA, Albillos A, Groszmann J. The role of central blood volume in the development of sodium retention in portal hypertensive rats. *Gastroenterology* 1996;110:193-198. DOI: 10.1053/gast.1996.v110.pm8536856

[162] Møller S, Wiinberg N, Henriksen JH. Noninvasive 24-hour ambulatory arterial blood pressure monitoring in cirrhosis. *Hepatology* 1995;22:88-95.

[163] Shapiro BP, Lam CS, Patel JB, Mohammed SF, Kruger M, Meyer DM, Linke WA, Redfield MM. Acute and chronic ventricular-arterial coupling in systole and diastole: insights from an elderly hypertensive model. *Hypertension* 2007;50(3):503-511. DOI: 10.1161/HYPERTENSIONAHA.107.090092

[164] Hansen TW, Li Y, Staessen JA. Blood pressure variability remains an elusive predictor of cardiovascular outcome. *Am J Hypertens* 2009;22(1):3-4. DOI: 10.1038/ajh.2008.322

[165] Fernandez-Varo G, Melgar-Lesmes P, Casals G, Pauta M, Arroyo V, Morales-Ruiz M, Ros J, Jimenez W. Inactivation of extrahepatic vascular Akt improves systemic hemodynamics and sodium excretion in cirrhotic rats. *J Hepatol* 2010;53(6):1041-1048. DOI: 10.1016/j.jhep.2010.05.031

[166] Møller S, Wiinberg N, Henriksen JH. Non-invasive 24-hour ambulatory arterial blood pressure monitoring in cirrhosis. *J Hepatol* 1992;16 (Suppl 1):S103.

[167] Henriksen JH, Fuglsang S, Bendtsen F. Arterial pressure profile in patients with cirrhosis: Fourier analysis of arterial pulse in relation to pressure level, stroke volume, and severity of disease: On the reduction of afterload in the hyperdynamic syndrome. *Scand J Gastroenterol* 2012;47(5):580-590. DOI: 10.3109/00365521.2012.658856

[168] Henriksen JH, Fuglsang S, Bendtsen F. Fourier analysis of arterial pulse in patients with advanced cirrhosis indicates reduced wave reflections that may protect against manifest cardiac dysfunction. *Gut* 2012;61(8):1237-1240. DOI: 10.1136/gutjnl-2011-301598

[169] Krag A, Borup T, Møller S, Bendtsen F. Efficacy and safety of terlipressin in cirrhotic patients with variceal bleeding or hepatorenal syndrome. *Adv Ther* 2008;25(11):1105-1140. DOI: 10.1007/s12325-008-0118-7

[170] Krag A, Bendtsen F, Pedersen EB, Holstein-Rathlou NH, Møller S. Effects of terlipressin on the aquaretic system - evidence of antidiuretic effects. *Am J Physiol Renal Physiol* 2008;295:F1295-F1300. DOI: 10.1152/ajprenal.90407.2008

[171] Pozzi M, Ratti L, Redaelli E, Guidi C, Mancia G. Cardiovascular abnormalities in special conditions of advanced cirrhosis. The circulatory adaptative changes to specific therapeutic procedures for the management of refractory ascites. *Gastroenterol Hepatol* 2006 29(4);263-272. DOI: 10.1157/13086820

[172] Wong F. Recent advances in our understanding of hepatorenal syndrome. *Nat Rev Gastroenterol Hepatol* 2012; 9(7):382-391.

[173] Tandon P, Garcia-Tsao G. Bacterial infections, sepsis, and multiorgan failure in cirrhosis. *Semin Liver Dis* 2008;28(1):26-42. DOI: 10.1055/s-2008-1040319

[174] Arroyo V, Fernandez J, Gines P. Pathogenesis and treatment of hepatorenal syndrome. *Semin Liver Dis* 2008;28(1):81-95. DOI: 10.1055/s-2008-1040323

[175] Lenz K. Hepatorenal syndrome--is it central hypovolemia, a cardiac disease, or part of gradually developing multiorgan dysfunction? *Hepatology* 2005;42(2):263-265. DOI: 10.1002/hep.20832

[176] Angeli P, Sanyal A, Møller S, Alessandria C, Gadano A, Kim R, Sarin SK, Bernardi M. Current limits and future challenges in the management of renal dysfunction in patients with cirrhosis: report from the International Club of Ascites. *Liver Int* 2013;33(1):16-23. DOI: 10.1111/j.1478-3231.2012.02807.x

[177] Iwakiri Y. The molecules: mechanisms of arterial vasodilatation observed in the splanchnic and systemic circulation in portal hypertension. *J Clin Gastroenterol* 2007;41(10 Suppl 3):S288-S294. DOI: 10.1097/MCG.0b013e3181468b4c

[178] Iwakiri Y. Endothelial dysfunction in the regulation of cirrhosis and portal hypertension. *Liver Int* 2012;32(2):199-213. DOI: 10.1111/j.1478-3231.2011.02579.x

[179] Abraldes JG, Iwakiri Y, Loureiro-Silva M, Haq O, Sessa WC, Groszmann RJ. Mild increases in portal pressure upregulate vascular endothelial growth factor and endothelial nitric oxide synthase in the intestinal microcirculatory bed, leading to a hyperdynamic state. *Am J Physiol Gastrointest Liver Physiol* 2006;G980-G987.

[180] Ruiz-Del-Arbol L, Monescillo A, Arocena C, Valer P, Gines P, Moreira V, Maria MJ, Jimenez W, Arroyo V. Circulatory function and hepatorenal syndrome in cirrhosis. *Hepatology* 2005;42:439-447. DOI: 10.1002/hep.20766

[181] Bernardi M, Maggioli C, Zaccherini G. Human albumin in the management of complications of liver cirrhosis. *Crit Care* 2012;20;16(2):211.

[182] Krag A, Bendtsen F, Burroughs AK, Møller S. The cardiorenal link in advanced cirrhosis. *Med Hypotheses* 2012;79(1):53-55. DOI: 10.1016/j.mehy.2012.03.032

[183] Rodriguez-Martinez M, Sawin LL, Dibona GF. Arterial and cardiopulmonary baroreflex control of renal nerve activity in cirrhosis. *Am J Physiol* 1995;268(1 Pt 2):R117-R129.

[184] Hendrickse MT, Triger DR. Vagal dysfunction and impaired urinary sodium and water excretion in cirrhosis. *Am J Gastroenterol* 1994;89(5):750-757.

[185] Gines P. Hepatorenal syndrome, pharmacological therapy, and liver transplantation. *Liver Transpl* 2011;17(11):1244-1246. DOI: 10.1002/lt.22433

[186] Lee SS, Marty J, Mantz J, Samain E, Braillon A, Lebrec D. Desensitization of myocardial beta-adrenergic receptors in cirrhotic rats. *Hepatology* 1990;12(3 Pt 1):481-485. DOI: 10.1002/hep.1840120306

[187] Tazi KA, Moreau R, Heller J, Poirel O, Lebrec D. Changes in protein kinase C isoforms in association with vascular hyporeactivity in cirrhotic rat aortas. *Gastroenterology* 2000;119(1):201-210. DOI: 10.1053/gast.2000.8522

[188] Theocharidou E, Krag A, Bendtsen F, Møller S, Burroughs AK. Cardiac dysfunction in cirrhosis - does adrenal function play a role? A hypothesis. *Liver Int* 2012;32(9):1327-1332. DOI: 10.1111/j.1478-3231.2011.02751.x

[189] Liu H, Gaskari SA, Lee SS. Cardiac and vascular changes in cirrhosis: pathogenic mechanisms. *World J Gastroenterol* 2006 Feb 14 2006;12:837-842.

[190] Dumcke CW, Møller S. Autonomic dysfunction in cirrhosis and portal hypertension. *Scand J Clin Lab Invest* 2008;68(6):437-447.

[191] Tandon P, Abraldes JG, Berzigotti A, Garcia-Pagan JC, Bosch J. Renin-angiotensin-aldosterone inhibitors in the reduction of portal pressure: A systematic review and meta-analysis. *J Hepatol* 2010;52(2):273-282 DOI: 10.1016/j.jhep.2010.03.013

[192] Bernardi M, Caraceni P. Pathogenesis of ascites and hepatorenal syndrome:altered haemodynamics and neurohumoral systems. In: Gerbes AL, Beuers U, Jungst D, Pape G, Sackmann M, Sauerbruch T, eds. *Hepatology* 2000. Falk symposium 117.Dordrecht: Kluwer Academic Publishers; 2001. p. 185-203.

[193] Sansoe G, Silvano S, Rosina F, Smedile A, Rizzetto M. Evidence of a dynamic aldosterone-independent distal tubular control of renal sodium excretion in compensated liver cirrhosis*. *J Intern Med* 2005;257(4):358-366. DOI: 10.1111/j.1365-2796.2005.01459.x

[194] Bernardi M, Santi L. Renal sodium retention in pre-ascitic cirrhosis: The more we know about the puzzle, the more it becomes intricate. *J Hepatol* 2010;53(5):790-792. DOI: 10.1016/j.jhep.2010.07.002

[195] La Villa G, Salmeron JM, Arroyo V, Bosch J, Gines P, Garcia-Pagan JC, Gines A, Asbert M, Jimenez W, Rivera F, Rodés J. Mineralocorticoid escape in patients with compensated cirrhosis and portal hypertension. *Gastroenterology* 1992;102:2114-2119.

[196] Zipprich A, Loureiro-Silva MR, Jain D, D'Silva I, Groszmann RJ. Nitric oxide and vascular remodeling modulate hepatic arterial vascular resistance in the isolated perfused cirrhotic rat liver. *J Hepatol* 2008;49:739-745. DOI: 10.1016/j.jhep.2008.06.027

[197] Guevara M, Gines P, Jimenez W, Sort P, Fernandezesparrach G, Escorsell A, Bataller R, Bosch J, Arroyo V, Rivera F, Rodes J, Fernandez-Esparrach G. Increased adrenomedullin levels in cirrhosis: Relationship with hemodynamic abnormalities and vasoconstrictor systems. *Gastroenterology* 1998;114:336-343. DOI: 10.1016/S0016-5085(98)70486-X

[198] Fabrega E, Casafont F, Crespo J, de la Pena J, San Miguel G, de las Heras G, Garcia-Unzueta MT, Amado JA, Pons-Romero F. Plasma adrenomedullin levels in patients with hepatic cirrhosis. *Am J Gastroenterol* 1997;92(10):1901-1904.

[199] Møller S, Henriksen JH. Circulatory abnormalities in cirrhosis with focus on neurohumoral aspects. *Semin Nephrol* 1997;17(6):505-519.

[200] Tahan V, Avsar E, Karaca C, Uslu E, Eren F, Aydin S, Uzun H, Hamzaoglu HO, Besisik F, Kalayci C, Okten A, Tozun N. Adrenomedullin in cirrhotic and non-cirrhotic portal hypertension. *World J Gastroenterol* 2003;9(10):2325-2327.

[201] Arroyo V, Fernandez J. Management of hepatorenal syndrome in patients with cirrhosis. *Nat Rev Nephrol* 2011;7(9):517-526.

[202] Davenport A, Cholongitas E, Xirouchakis E, Burroughs AK. Pitfalls in assessing renal function in patients with cirrhosis-potential inequity for access to treatment of hepatorenal failure and liver transplantation. *Nephrol Dial Transplant* 2011;26(9):2735-2742. DOI: 10.1093/ndt/gfr354

[203] Huisman EJ, Trip EJ, Siersema PD, van Hoek B, van Erpecum KJ. Protein energy malnutrition predicts complications in liver cirrhosis. *Eur J Gastroenterol Hepatol* 2011;23(11):982-989. DOI: 10.1097/MEG.0b013e32834aa4bb

[204] Bernardi M, Trevisani F, Gasbarrini A, Gasbarrini G. Hepatorenal disorders.Role of the renin-angiotensin-aldosterone system. *Sem Liver Dis* 1994;14:23-34. DOI: 10.1055/s-2007-1007295

[205] Fernandez-Seara J, Prieto J, Quiroga J, Zozaya JM, Cobos MA, Rodriguez-Eire JL, Garzia-Plaza A, Leal J. Systemic and regional hemodynamics in patients with liver cirrhosis and ascites with and without functional renal failure. *Gastroenterology* 1989;97:1304-1312.

[206] Moore K. Acute kidney injury in cirrhosis - a changing spectrum. *Hepatology* 2012;57 :435-437. DOI: 10.1002/hep.26003

[207] Henriksen JH, Ring-Larsen H. Hepatorenal disorders:Role of the sympathetic nervous system. *Sem Liver Dis* 1994;14:35-43. DOI: 10.1055/s-2007-1007296

[208] Ring-Larsen H. Renal blood flow in cirrhosis: relation to systemic and portal haemodynamics and liver function. *Scand J Clin Lab Invest* 1977;37:635-642. DOI: 10.3109/00365517709100657

[209] Claria J, Rodes J. Renal sodium handling in preascitic cirrhosis. Gut 2001;48(5):740-741. DOI: 10.1136/gut.48.5.740a

[210] Levy M. Pathogenesis of sodium retention in early cirrhosis of the liver:Evidence for vascular overfilling. *Sem Liver Dis* 1994;14:4-13. DOI: 10.1055/s-2007-1007293

[211] Bernardi M, Li BS, Arienti V, de Collibus C, Scialpi C, Boriani L, Zanzani S, Caraceni P, Trevisani F. Systemic and regional hemodynamics in pre-ascitic cirrhosis: effects of posture. *J Hepatol* 2003;39(4):502-508. DOI: 10.1016/S0168-8278(03)00324-6

[212] Moore KP, Aithal GP. Guidelines on the management of ascites in cirrhosis. *Gut* 2006;55 Suppl 6:vi1-vi12. DOI: 10.1136/gut.2006.099580

[213] Salerno F, Gerbes A, Gines P, Wong F, Arroyo V. Diagnosis, prevention and treatment of hepatorenal syndrome in cirrhosis. *Postgrad Med J* 2008;84(998):662-670. DOI: 10.1136/gut.2006.107789

[214] Dibona GF, Kopp UC. Neural control of renal function. *Physiol Rev* 1997;77(1):75-197.

[215] Moore K, Wendon J, Frazer M, Karani J, Williams R, Badr K. Plasma endothelin immunoreactivity in liver disease and the hepatorenal syndrome. *N Engl J Med* 1992;327:1774-1778. DOI: 10.1056/NEJM199212173272502

[216] Capella GL. Anti-leukotriene drugs in the prevention and treatment of hepatorenal syndrome. *Prostaglandins Leukot Essent Fatty Acids* 2003;68(4):263-265. DOI: 10.1016/S0952-3278(03)00004-8

[217] Yu Z, Serra A, Sauter D, Loffing J, Ackermann D, Frey FJ, Frey BM, Vogt B. Sodium retention in rats with liver cirrhosis is associated with increased renal abundance of NaCl cotransporter (NCC). *Nephrol Dial Transplant* 2005;20(9):1833-1841. DOI: 10.1093/ndt/gfh916

[218] Henriksen JH, Ring-Larsen H. Renal effects of drugs used in the treatment of portal hypertension. *Hepatology* 1993;18:688-695. DOI: 10.1002/hep.1840180329

[219] Ottesen LH, Aagaard NK, Kiszka-Kanowitz M, Rehling M, Henriksen JH, Pedersen EB, Flyvbjerg A, Bendtsen F. Effects of a long-acting formulation of octreotide on renal function and renal sodium handling in cirrhotic patients with portal hypertension: A randomized, double-blind, controlled trial. *Hepatology* 2001;34(3):471-477. DOI: 10.1053/jhep.2001.26754

[220] Martin-Llahi M, Pepin MN, Guevara M, Diaz F, Torre A, Monescillo A, Soriano G, Terra C, Fabrega E, Arroyo V, Rodes J, Gines P. Terlipressin and albumin vs albumin in patients with cirrhosis and hepatorenal syndrome: a randomized study. *Gastroenterology* 2008;134(5):1352-1359. DOI: 10.1053/j.gastro.2008.02.024

[221] Angeli P, Gatta A, Caregaro L, Menon F, Sacerdoti D, Merkel C, Rondana M, de Toni R, Ruol A. Tubular site of renal sodium retention in ascitic liver cirrhosis evaluated by lithium clearance. *Eur J Clin Invest* 1990;20(1):111-117. DOI: 10.1111/j.1365-2362.1990.tb01800.x

[222] Krag A, Møller S, Pedersen EB, Henriksen JH, Holstein-Rathlou NH, Bendtsen F. Impaired free water excretion in child C cirrhosis and ascites: relations to distal tubular function and the vasopressin system. *Liver Int* 2010;30(9):1364-1370. DOI: 10.1111/j.1478-3231.2010.02319.x

[223] Jonassen TE, Christensen S, Kwon TH, Langhoff S, Salling N, Nielsen S. Renal water handling in rats with decompensated liver cirrhosis. *Am J Physiol Renal Physiol* 2000;279(6):F1101-F1109.

[224] Sansoe G, Ferrari A, Baraldi E, Castellana CN, De Santis MC, Manenti F. Renal distal tubular handling of sodium in central fluid volume homoeostasis in preascitic cirrhosis. *Gut* 1999;45(5):750-755. DOI: 10.1136/gut.45.5.750

[225] Schrier RW, Masoumi A, Elhassan E. Aldosterone: role in edematous disorders, hypertension, chronic renal failure, and metabolic syndrome. *Clin J Am Soc Nephrol* 2010;5(6):1132-1140. DOI: 10.2215/CJN.01410210

[226] Ackermann D, Mordasini D, Cheval L, Imbert-Teboul M, Vogt B, Doucet A. Sodium retention and ascites formation in a cholestatic mice model: role of aldosterone and mineralocorticoid receptor? *Hepatology* 2007;46(1):173-179. DOI: 10.1002/hep.21699

[227] Thiesson HC, Jensen BL, Bistrup C, Ottosen PD, McNeilly AD, Andrew R, Seckl J, Skott O. Renal sodium retention in cirrhotic rats depends on glucocorticoid-mediated activation of the mineralocorticoid receptor due to decreased renal 11{beta}HSD2 activity. *Am J Physiol Regul Integr Comp Physiol* 2006;292(1):R625-R636. DOI: 10.1152/ajpregu.00418.2005

[228] Sansoe G, Aragno M, Tomasinelli CE, di Bonzo LV, Wong F, Parola M. Calcium-dependent diuretic system in preascitic liver cirrhosis. *J Hepatol* 2010;53(5):856-862. DOI: 10.1016/j.jhep.2010.05.021

[229] Jonassen TEN, Marcussen N, Haugan K, Skyum H, Christensen S, Andreasen F, Petersen JS. Functional and structural changes in the thick ascending limb of Henle's loop in rats with liver cirrhosis. *Amer J Physiol-Regul Integr C* 1998;43:CP3-CP3.

[230] Krag A, Pedersen EB, Møller S, Bendtsen F. Effects of the vasopressin agonist terlipressin on plasma cAMP and ENaC excretion in the urine in patients with cirrhosis and water retention. *Scand J Clin Lab Invest* 2010;71(2):112-116.

[231] Radin MJ, Yu MJ, Stødkilde L, Lance MR, Hoffert JD, Frøkiær J, Pisitkun T, Knepper MA. Aquaporin-2 regulation in health and disease. *Vet Clin Pathol* 2012;41(4):455-470. DOI: 10.1111/j.1939-165x.2012.00488.x

[232] Jonassen TEN, Promeneur D, Christensen S, Petersen JS, Nielsen S. Decreased vasopressin-mediated renal water reabsorption in rats with chronic aldosterone-receptor blockade. *Amer J Physiol Renal Physiol* 2000;278(2):F246-F256.

[233] Fabrega E, Berja A, Garcia-Unzueta MT, Guerra-Ruiz A, Cobo M, Lopez M, Bolado-Carrancio A, Amado JA, Rodriguez-Rey JC, Pons-Romero F. Influence of aquaporin-1 gene polymorphism on water retention in liver cirrhosis. *Scand J Gastroenterol* 2011;46(10):1267-1274. DOI: 10.3109/00365521.2011.603161

[234] Schrier RW. Vasopressin and aquaporin 2 in clinical disorders of water homeostasis. *Semin Nephrol* 2008;28(3):289-296. DOI: 10.1016/j.semnephrol.2008.03.009

[235] Gines P, Wong F, Watson H, Terg R, Bruha R, Zarski JP. Clinical trial: short-term effects of combination of satavaptan, a selective vasopressin V receptor antagonist, and diuretics on ascites in patients with cirrhosis without hyponatremia - a randomized, double-blind, placebo-controlled study. *Aliment Pharmacol Ther* 2010;31(8);834-845.

[236] Schrier RW, Sharma S, Shchekochikhin D. Hyponatraemia: more than just a marker of disease severity? *Nat Rev Nephrol* 2013;2(1):37-50.

[237] Wong F, Watson H, Gerbes A, Vilstrup H, Badalamenti S, Bernardi M, Gines P. Satavaptan for the management of ascites in cirrhosis: efficacy and safety across the spectrum of ascites severity. *Gut* 2012;61(1):108-116. DOI: 10.1136/gutjnl-2011-300157

[238] Ivarsen P, Frokiaer J, Aagaard NK, Hansen EF, Bendtsen F, Nielsen S, Vilstrup H. Increased urinary excretion of aquaporin 2 in patients with liver cirrhosis. *Gut* 2003;52(8):1194-1199. DOI: 10.1136/gut.52.8.1194

[239] Esteva-Font C, Baccaro ME, Fernandez-Llama P, Sans L, Guevara M, Ars E, Jimenez W, Arroyo V, Ballarin JA, Gines P. Aquaporin-1 and aquaporin-2 urinary excretion in cirrhosis: Relationship with ascites and hepatorenal syndrome. *Hepatology* 2006;44(6):1555-1563. DOI: 10.1002/hep.21414

[240] Gines P, Wong F, Watson H, Milutinovic S, del Arbol LR, Olteanu D. Effects of satavaptan, a selective vasopressin V(2) receptor antagonist, on ascites and serum sodium in cirrhosis with hyponatremia: a randomized trial. *Hepatology* 2008;48(1):204-213. DOI: 10.1002/hep.22293

[241] Gines P. Vaptans: A promising therapy in the management of advanced cirrhosis. *J Hepatol* 2007;46(6):1150-1152. DOI: 10.1016/j.jhep.2007.03.007

[242] Møller S, Henriksen JH. Pathogenesis and pathophysiology of hepatorenal syndrome - is there scope for prevention? *Aliment Pharmacol Ther* 2004;20 Suppl 3:31-41.

[243] Jalan R, Forrest EH, Redhead DN, Dillon JF, Hayes PC. Reduction in renal blood flow following acute increase in the portal pressure: Evidence for the existence of a hepatorenal reflex in man? *Gut* 1997;40:664-670.

[244] Jimenez-Saenz M, Soria IC, Bernardez JR, Gutierrez JM. Renal sodium retention in portal hypertension and hepatorenal reflex: from practice to science. *Hepatology* 2003;37(6):1494-1495. DOI: 10.1053/jhep.2003.50226

[245] Lautt WW. Regulatory processes interacting to maintain hepatic blood flow constancy: Vascular compliance, hepatic arterial buffer response, hepatorenal reflex, liver regeneration, escape from vasoconstriction. *Hepatol Res* 2007;37(11):891-903. DOI: 10.1111/j.1872-034X.2007.00148.x

[246] Kiil F. The mechanism of renal autoregulation. *Scand J Clin Lab Invest* 1981;41(6):521-525. DOI: 10.3109/00365518109090493

[247] Piano S, Morando F, Fasolato S, Cavallin M, Boscato N, Boccagni P, Zanus G, Cillo U, Gatta A, Angeli P. Continuous recurrence of type 1 hepatorenal syndrome and long-term treatment with terlipressin and albumin: a new exception to meld score in the allocation system to liver transplantation? *J Hepatol* 2011;55(2):491-496. DOI: 10.1016/j.jhep.2011.02.002

[248] Krag A, Møller S. Safety of terlipressin for hepatorenal syndrome. In: Gerbes A, ed. *Ascites, hyponatremia and hepatrenal syndrome: Progress in treatment*. 1 ed. Basel: Karger; 2011. p. 178-188.

[249] Cardenas A, Gines P, Marotta P, Czerwiec F, Oyuang J, Guevara M, Afdhal NH. Tolvaptan, an oral vasopressin antagonist, in the treatment of hyponatremia in cirrhosis. *J Hepatol* 2012;56(3):571-578. DOI: 10.1016/j.jhep.2011.08.020

[250] Mani AR, Ippolito S, Ollosson R, Moore KP. Nitration of cardiac proteins is associated with abnormal cardiac chronotropic responses in rats with biliary cirrhosis. *Hepatology* 2006;43:847-856. DOI: 10.1002/hep.21115

[251] Babaei-Karamshahlou M, Hooshmand B, Hajizadeh S, Mani AR. The role of endogenous hydrogen sulfide in pathogenesis of chronotropic dysfunction in rats with cirrhosis. *Eur J Pharmacol* 2012;696(1-3):130-135. DOI: 10.1016/j.ejphar.2012.09.039

[252] Ros J, Claria J, To-Figueras J, Planaguma A, Cejudo-Martin P, Fernandez-Varo G, Martin-Ruiz R, Arroyo V, Rivera F, Rodes J, Jimenez W. Endogenous cannabinoids: a new system involved in the homeostasis of arterial pressure in experimental cirrhosis in the rat. *Gastroenterology* 2002;122(1):85-93. DOI: 10.1053/gast.2002.30305

[253] Caraceni P, Viola A, Piscitelli F, Giannone F, Berzigotti A, Cescon M, Domenicali M, Petrosino S, Giampalma E, Riili A, Grazi G, Golfieri R, Zoli M, Bernardi M, Di M, V. Circulating and hepatic endocannabinoids and endocannabinoid-related molecules in patients with cirrhosis. *Liver Int* 2010;30(6);816-825. DOI: 10.1111/j.1478-3231.2009.02137.x

[254] Mallat A, Teixeira-Clerc F, Deveaux V, Manin S, Lotersztajn S. The endocannabinoid system as a key mediator during liver diseases: New insights and therapeutic openings. *Br J Pharmacol* 2011;163(7):1432-1440. DOI: 10.1111/j.1476-5381.2011.01397.x

[255] Pacher P, Gao B. Endocannabinoids and liver disease. III. Endocannabinoid effects on immune cells: implications for inflammatory liver diseases. *Am J Physiol Gastrointest Liver Physiol* 2008;294(4):G850-G854. DOI: 10.1152/ajpgi.00523.2007

[256] Gaskari SA, Liu H, Moezi L, Li Y, Baik SK, Lee SS. Role of endocannabinoids in the pathogenesis of cirrhotic cardiomyopathy in bile duct-ligated rats. *Br J Pharmacol* 2005;146(3):315-323. DOI: 10.1038/sj.bjp.0706331

[257] Yang YY, Tsai TH, Huang YT, Lee TY, Chan CC, Lee KC, Lin HC. Hepatic endothelin-1 and endocannabinoids-dependent effects of hyperleptinaemia in nonalcoholic steatohepatitis-cirrhotic rats. *Hepatology* 2012;55(5):1540-1550. DOI: 10.1002/hep.25534

[258] Lin HC, Yang YY, Tsai TH, Huang CM, Huang YT, Lee FY, Liu TT, Lee SD. The relationship between endotoxemia and hepatic endocannabinoids in cirrhotic rats with portal hypertension. *J Hepatol* 2011;54(6):1145-1153. DOI: 10.1016/j.jhep.2010.09.026

[259] Wong F, Girgrah N, Graba J, Allidina Y, Liu P, Blendis L. The cardiac response to exercise in cirrhosis. *Gut* 2001;49(2):268-275. DOI: 10.1136/gut.49.2.268

[260] Møller S, Henriksen JH. Cardiovascular dysfunction in cirrhosis. Pathophysiological evidence of a cirrhotic cardiomyopathy. *Scand J Gastroenterol* 2001;36(8):785-794. DOI: 10.1080/003655201750313289

[261] Krag A, Bendtsen F, Mortensen C, Henriksen JH, Møller S. Effects of a single terlipressin administration on cardiac function and perfusion in cirrhosis. *Eur J Gastroenterol Hepatol* 2010;22(9):1085-1092. DOI: 10.1097/MEG.0b013e32833a4822

[262] Huonker M, Schumacher YO, Ochs A, Sorichter S, Keul J, Rôssle M. Cardiac function and haemodynamics in alcoholic cirrhosis and effects of the transjugular intrahepatic portosystemic stent shunt. *Gut* 1999;44(5):743-748. DOI: 10.1136/gut.44.5.743

[263] Merli M, Valeriano V, Funaro S, Attili AF, Masini A, Efrati C, De CS, Riggio O. Modifications of cardiac function in cirrhotic patients treated with transjugular intrahepatic portosystemic shunt (TIPS). *Am J Gastroenterol* 2002;97(1):142-148. DOI: 10.1111/j.1572-0241.2002.05438.x

[264] Gines P, Uriz J, Calahorra B, Garcia-Tsao G, Kamath PS, del Arbol LR, Planas R, Bosch J, Arroyo V, Rodes J. Transjugular intrahepatic portosystemic shunting versus paracentesis plus albumin for refractory ascites in cirrhosis. *Gastroenterology* 2002;123(6):1839-1847. DOI: 10.1053/gast.2002.37073

[265] Møller S, Henriksen JH. Cardiovascular dysfunction in cirrhosis. Pathophysiological evidence of a cirrhotic cardiomyopathy. *Scand J Gastroenterol* 2001;36(8):785-794. DOI: 10.1080/003655201750313289

[266] Pozzi M, Carugo S, Boari G, Pecci V, de Ceglia S, Maggiolini S, Bolla GB, Roffi L, Failla M, Grassi G, Giannattasio C, Mancia G. Evidence of functional and structural cardiac

abnormalities in cirrhotic patients with and without ascites. *Hepatology* 1997;26(5):1131-1137.

[267] Torregrosa M, Aguade S, Dos L, Segura R, Gonzalez A, Evangelista A, Castell J, Margarit C, Esteban R, Guardia J, Genesca J. Cardiac alterations in cirrhosis: reversibility after liver transplantation. *J Hepatol* 2005;42(1):68-74. DOI: 10.1016/j.jhep.2004.09.008

[268] Grose RD, Nolan J, Dillon JF, Errington M, Hannan WJ, Bouchier IAD, Hayes PC. Exercise-induced left ventricular dysfunction in alcoholic and non-alcoholic cirrhosis. *J Hepatol* 1995;22:326-332. DOI: 10.1016/0168-8278(95)80286-X

[269] Epstein SK, Ciubotaru RL, Zilberberg MD, Kaplan LM, Jacoby C, Freeman R, Kaplan MM. Analysis of impaired exercise capacity in patients with cirrhosis. *Dig Dis Sci* 1998;43(8):1701-1707. DOI: 10.1023/A:1018867232562

[270] Pozzi M, Redaelli E, Ratti L, Poli G, Guidi C, Milanese M, Calchera I, Mancia G. Time-course of diastolic dysfunction in different stages of chronic HCV related liver diseases. *Minerva Gastroenterol Dietol* 2005;51(2):179-186. DOI:

[271] Gaskari SA, Honar H, Lee SS. Therapy insight: Cirrhotic cardiomyopathy. *Nat Clin Pract Gastroenterol Hepatol* 2006;3(6):329-337. DOI: 10.1038/ncpgasthep0498

[272] Saner FH, Neumann T, Canbay A, Treckmann JW, Hartmann M, Goerlinger K, Bertram S, Beckebaum S, Cicinnati V, Paul A. High brain-natriuretic peptide level predicts cirrhotic cardiomyopathy in liver transplant patients. *Transpl Int* 2011;24:425-432. DOI: 10.1111/j.1432-2277.2011.01219.x

[273] Finucci G, Desideri A, Sacerdoti D, Bolognesi M, Merkel C, Angeli P, Gatta A. Left ventricular diastolic function in liver cirrhosis. *Scand J Gastroenterol* 1996;31 :279-284. DOI: 10.3109/00365529609004879

[274] Rabie RN, Cazzaniga M, Salerno F, Wong F. The use of E/A ratio as a predictor of outcome in cirrhotic patients treated with transjugular intrahepatic portosystemic shunt. Am *J Gastroenterol* 2009;104:2458-2466. DOI: 10.1038/ajg.2009.321

[275] Cazzaniga M, Salerno F, Pagnozzi G, Dionigi E, Visentin S, Cirello I, Meregaglia D, Nicolini A. Diastolic dysfunction is associated with poor survival in cirrhotic patients with transjugular intrahepatic portosystemic shunt. *Gut* 2007;56(6):869-875. DOI: 10.1136/gut.2006.102467

[276] Nazar A, Guevara M, Sitges M, Terra C, Sola E, Guigou C, Arroyo V, Gines P. Left ventricular function assessed by echocardiography in cirrhosis: Relationship to systemic hemodynamics and renal dysfunction. *J Hepatol* 2013;58(1):51-57. DOI: 10.1016/j.jhep.2012.08.027

[277] Dowsley TF, Bayne DB, Langnas AN, Dumitru I, Windle JR, Porter TR, Raichlin E. Diastolic dysfunction in patients with end-stage liver disease is associated with development of heart failure early after liver transplantation. *Transplantation* 2012;94(6);646-651. DOI: 10.1097/TP.0b013e31825f0f97

[278] Pozzi M, Grassi G, Ratti L, Favini G, Dell'Oro R, Redaelli E, Calchera I, Boari G, Mancia G. Cardiac, neuroadrenergic, and portal hemodynamic effects of prolonged aldosterone blockade in postviral child a cirrhosis. *Am J Gastroenterol* 2005;100 (5):1110-1116. DOI: 10.1111/j.1572-0241.2005.41060.x

[279] Bos R, Mougenot N, Findji L, Mediani O, Vanhoutte PM, Lechat P. Inhibition of Catecholamine-Induced Cardiac Fibrosis by an Aldosterone Antagonist. *J Cardiovasc Pharmacol* 2005; 45(1):8-13. DOI: 10.1097/00005344-200501000-00003

[280] Myers RP, Lee SS. Cirrhotic cardiomyopathy and liver transplantation. *Liver Transpl* 2000;6(4 Suppl 1):S44-S52. DOI: 10.1002/lt.500060510

[281] Karabulut A, Iltumur K, Yalcin K, Toprak N. Hepatopulmonary syndrome and right ventricular diastolic functions: an echocardiographic examination. *Echocardiography* 2006;23:271-278. DOI: 10.1111/j.1540-8175.2006.00210.x

[282] Pouriki S, Alexopoulou A, Chrysochoou C, Raftopoulos L, Papatheodoridis G, Stefanadis C, Pectasides D. Left ventricle enlargement and increased systolic velocity in the mitral valve are indirect markers of the hepatopulmonary syndrome. *Liver Int* 2011;31:1388-1394. DOI: 10.1111/j.1478-3231.2011.02591.x

[283] Henriksen JH, Fuglsang S, Bendtsen F, Christensen E, Møller S. Dyssynchronous electrical and mechanical systole in patients with cirrhosis. *J Hepatol* 2002;36(4):513-520. DOI: 10.1016/S0168-8278(02)00010-7

[284] Zambruni A, Trevisani F, Caraceni P, Bernardi M. Cardiac electrophysiological abnormalities in patients with cirrhosis. *J Hepatol* 2006 2006;44:994-1002. DOI: 10.1016/j.jhep.2005.10.034

[285] Trevisani F, Di Micoli A, Zambruni A, Biselli M, Santi V, Erroi V, Lenzi B, Caraceni P, Domenicali M, Cavazza M, Bernardi M. QT interval prolongation by acute gastrointestinal bleeding in patients with cirrhosis. *Liver Int* 2012;32(10):1510-1515.

[286] Zavecz JH, Bueno O, Maloney RE, ODonnell JM, Roerig SC, Battarbee HD. Cardiac excitation-contraction coupling in the portal hypertensive rat. *Amer J Physiol Gastrointest L* 2000;279(1):G28-G39.

[287] Henriksen JH, Fuglsang S, Bendtsen F, Christensen E, Møller S. Arterial compliance in patients with cirrhosis. High stroke volume/pulse pressure ratio as an index of elevated arterial compliance. *Am J Physiol* 2001;280:G584-G594.

[288] Ward CA, Ma Z, Lee SS, Giles WR. Potassium currents in atrial and ventricular myocytes from a rat model of cirrhosis. *Am J Physiol* 1997;273:G537-G544.

[289] Henriksen JH, Bendtsen F, Hansen EF, Møller S. Acute non-selective beta-adrenergic blockade reduces prolonged frequency-adjusted Q-T interval (QTc) in patients with cirrhosis. *J Hepatol* 2004;40(2):239-246. DOI: 10.1016/j.jhep.2003.10.026

[290] Henriksen JH, Gøetze JP, Fuglsang S, Christensen E, Bendtsen F, Møller S. Increased circulating pro-brain natriuretic peptide (proBNP) and brain natriuretic peptide (BNP) in patients with cirrhosis: relation to cardiovascular dysfunction and severity of disease. *Gut* 2003;52(10):1511-1517. DOI: 10.1136/gut.52.10.1511

[291] Guazzi M, Polese A, Magrini F, Fiorentini C, Olivari MT. Negative influences of ascites on the cardiac function of cirrhotic patients. *Am J Med* 1975;59:165-170. DOI: 10.1016/0002-9343(75)90350-2

[292] Fernandez J, Acevedo J, Castro M, Garcia O, Rodriguez dL, Roca D, Pavesi M, Sola E, Moreira L, Silva A, Seva-Pereira T, Corradi F, Mensa J, Gines P, Arroyo V. Prevalence and risk factors of infections by multiresistant bacteria in cirrhosis: A prospective study. *Hepatology* 2012;55(5):1551-1561. DOI: 10.1002/hep.25532

[293] Wiest R, Krag A, Gerbes A. Spontaneous bacterial peritonitis: recent guidelines and beyond. *Gut* 2012;61(2):297-310. DOI: 10.1136/gutjnl-2011-300779

[294] Evans LT, Kim WR, Poterucha JJ, Kamath PS. Spontaneous bacterial peritonitis in asymptomatic outpatients with cirrhotic ascites. *Hepatology* 2003;37(4):897-901. DOI: 10.1053/jhep.2003.50119

[295] Fernandez J, Gustot T. Management of bacterial infections in cirrhosis. *J Hepatol* 2012;56 Suppl 1:S1-12.

[296] Bendtsen F, Grønbæk H, Hansen JB, Aagaard NK, Schmidt L, Møller S. Treatment of ascites and spontaneous bacterial peritonitis - Part I. *Dan Med J* 2012;59(1):C4371.

[297] Hung TH, Tsai CC, Hsieh YH, Tsai CC, Tseng CW, Tsai JJ. Effect of renal Impairment on mortality of patients with cirrhosis and spontaneous bacterial peritonitis. *Clin Gastroenterol Hepatol* 2012;10(6):677-681. DOI: 10.1016/j.cgh.2012.02.026

[298] Ring-Larsen H, Henriksen JH, Wilken C, Clausen J, Pals H, Christensen NJ. Diuretic treatment in decompensated cirrhosis and congestive heart failure: effect of posture. *Br Med J* (Clin Res Ed) 1986;292(6532):1351-1353. DOI: 10.1136/bmj.292.6532.1351

[299] Gerbes AL, Gulberg V, Gines P, Decaux G, Gross P, Gandjini H, Djian J. Therapy of hyponatremia in cirrhosis with a vasopressin receptor antagonist: A randomized double-blind multicenter trial. *Gastroenterology* 2003;124(4):933-939. DOI: 10.1053/gast.2003.50143

[300] Gines P, Guevara M. Hyponatremia in cirrhosis: pathogenesis, clinical significance, and management. *Hepatology* 2008;48(3):1002-1010. DOI: 10.1002/hep.22418

[301] Fernandez-Esparrach G, Guevara M, Sort P, Pardo A, Jimenez W, Gines P, Planas R, Lebrec D, Geuvel A, Elewaut A, Adler M, Arroyo V. Diuretic requirements after therapeutic paracentesis in non-azotemic patients with cirrhosis. A randomized double-blind trial of spironolactone versus placebo. *J Hepatol* 1997;26(3):614-620. DOI: 10.1016/S0168-8278(97)80427-8

[302] Cabrera J, Falcon L, Gorriz E, Pardo MD, Granados R, Quinones A, Maynar M. Abdominal decompression plays a major role in early postparacentesis haemodynamic changes in cirrhotic patients with tense ascites. *Gut* 2001;48(3):384-389. DOI: 10.1136/gut.48.3.384

[303] Salerno F, Guevara M, Bernardi M, Moreau R, Wong F, Angeli P, Garcia-Tsao G, Lee SS. Refractory ascites: pathogenesis, definition and therapy of a severe complication in patients with cirrhosis. *Liver Int* 2010;30(7):937-947. DOI: 10.1111/j.1478-3231.2010.02272.x

[304] Salerno F, Badalamenti S, Lorenzano E, Moser P, Incerti P. Randomized comparative study of hemaccel vs. albumin infusion after total paracentesis in cirrhotic patients with refractory ascites. *Hepatology* 1991;13(4):707-713. DOI: 10.1002/hep.1840130416

[305] Fassio E, Terg R, Landeira G, Abecasis R, Salemne M, Podesta A, Rodriguez P, Levi D, Kravetz D. Paracentesis with Dextran 70 vs. paracentesis with albumin in cirrhosis with tense ascites. Results of a randomized study. *J Hepatol* 1992;14(2-3):310-316. DOI: 10.1016/0168-8278(92)90176-P

[306] Altman C, Bernard B, Roulot D, Vitte RL, Ink O. Randomized comparative multicenter study of hydroxyethyl starch versus albumin as a plasma expander in cirrhotic patients with tense ascites treated with paracentesis. *Eur J Gastroenterol Hepatol* 1998;10(1):5-10. DOI: 10.1097/00042737-199801000-00002

[307] Planas R, Gines P, Arroyo V, Llach J, Panes J, Vargas V, Salmeron JM, Gines A, Toledo C, Rimola A, Jimenez W, Asbert M, Gassull A, Rodes J. Dextran-70 versus albumin as plasma expanders in cirrhotic patients with tense ascites treated with total paracentesis. *Gastroenterology* 1990;99:1736-1744.

[308] Gines A, Fernandez-Esparrach G, Monescillo A, Vila C, Domenech E, Abecasis R, Angeli P, Ruiz-Del-Arbol L, Planas R, Rodes J. Randomized trial comparing albumin, dextran 70, and polygeline in cirrhotic patients with ascites treated by paracentesis. *Gastroenterology* 1996;111:1002-1010. DOI: 10.1016/S0016-5085(96)70068-9

[309] Moreau R, Valla DC, Durand-Zaleski I, Bronowicki JP, Durand F, Chaput JC, Dadamessi I, Silvain C, Bonny C, Oberti F, Gournay J, Lebrec D, Grouin JM, Guemas E, Golly D, Padrazzi B, Tellier Z. Comparison of outcome in patients with cirrhosis and ascites following treatment with albumin or a synthetic colloid: a randomised controlled pilot trail. *Liver Int* 2006;26(1):46-54. DOI: 10.1111/j.1478-3231.2005.01188.x

[310] Moreau R, Asselah T, Condat B, de Kerguenec C, Pessione F, Bernard B, Poynard T, Binn M, Grange J, Valla D, Lebrec D. Comparison of the effect of terlipressin and albumin on arterial blood volume in patients with cirrhosis and tense ascites treated by paracentesis: a randomised pilot study. *Gut* 2002;50(1):90-94. DOI: 10.1136/gut.50.1.90

[311] Singh V, Kumar B, Nain CK, Singh B, Sharma N, Bhalla A, Sharma AK. Noradrenaline and albumin in paracentesis-induced circulatory dysfunction in cirrhosis: a randomized pilot study. *J Intern Med* 2006;260(1):62-68. DOI: 10.1111/j.1365-2796.2006.01654.x

[312] Singh V, Dheerendra PC, Singh B, Nain CK, Chawla D, Sharma N, Bhalla A, Mahi SK. Midodrine versus albumin in the prevention of paracentesis-induced circulatory dysfunction in cirrhotics: a randomized pilot study. *Am J Gastroenterol* 2008;103(6):1399-1405. DOI: 10.1111/j.1572-0241.2008.01787.x

[313] Appenrodt B, Wolf A, Grunhage F, Trebicka J, Schepke M, Rabe C, Lammert F, Sauerbruch T, Heller J. Prevention of paracentesis-induced circulatory dysfunction: midodrine vs albumin. A randomized pilot study. *Liver Int* 2008;28(7):1019-1025. DOI: 10.1111/j.1478-3231.2008.01734.x

[314] D'amico G, Luca A, Morabito A, Miraglia R, D'Amico M. Uncovered transjugular intrahepatic portosystemic shunt for refractory ascites: a meta-analysis. *Gastroenterology* 2005;129(4):1282-1293. DOI: 10.1053/j.gastro.2005.07.031

[315] Rossle M, Ochs A, Gulberg V, Siegerstetter V, Holl J, Deibert P, Olschewski M, Reiser M, Gerbes AL. A comparison of paracentesis and transjugular intrahepatic portosystemic

shunting in patients with ascites. *N Engl J Med* 2000;342(23):1701-1707. DOI: 10.1056/NEJM200006083422303

[316] Salerno F, Camma C, Enea M, Rossle M, Wong F. Transjugular intrahepatic portosystemic shunt for refractory ascites: a meta-analysis of individual patient data. *Gastroenterology* 2007;133(3):825-834. DOI: 10.1053/j.gastro.2007.06.020

[317] Saab S, Nieto JM, Lewis SK, Runyon BA. TIPS versus paracentesis for cirrhotic patients with refractory ascites. *Cochrane Database Syst Rev* 2006;(4):CD004889.

[318] Bureau C, Garcia-Pagan JC, Otal P, Pomier-Layrargues G, Chabbert V, Cortez C, Perreault P, Peron JM, Abraldes JG, Bouchard L, Bilbao JI, Bosch J, Rousseau H, Vinel JP. Improved clinical outcome using polytetrafluoroethylene-coated stents for TIPS: results of a randomized study. *Gastroenterology* 2004;126(2):469-475. DOI: 10.1053/j.gastro.2003.11.016

[319] Sort P, Navasa M, Arroyo V, Aldeguer X, Planas R, Ruizdelarbol L, Castells L, Vargas V, Soriano G, Guevara M, Gines P, Rodes J. Effect of intravenous albumin on renal impairment and mortality in patients with cirrhosis and spontaneous bacterial peritonitis. *N Engl J Med* 1999;341(6):403-409. DOI: 10.1056/NEJM199908053410603

[320] Andreu M, Sola R, Sitges-Serra A, Alia C, Gallen M, Vila MC, Coll S, Oliver MI. Risk factors for spontaneous bacterial peritonitis in cirrhotic patients with ascites. *Gastroenterology* 1993;104(4):1133-1138.

[321] Tandon P, Garcia-Tsao G. Renal dysfunction is the most important independent predictor of mortality in cirrhotic patients with spontaneous bacterial peritonitis. *Clin Gastroenterol Hepatol* 2010;2011,9(3):260-265.

[322] Terg R, Fassio E, Guevara M, Cartier M, Longo C, Lucero R, Landeira C, Romero G, Dominguez N, Munoz A, Levi D, Miguez C, Abecasis R. Ciprofloxacin in primary prophylaxis of spontaneous bacterial peritonitis: a randomized, placebo-controlled study. *J Hepatol* 2008;48(5):774-779. DOI: 10.1016/j.jhep.2008.01.024

[323] Gines P, Rimola A, Planas R, Vargas V, Marco F, Almela M, Forne M, Miranda ML, Llach J, Salmeron JM, . Norfloxacin prevents spontaneous bacterial peritonitis recurrence in cirrhosis: results of a double-blind, placebo-controlled trial. *Hepatology 1990;12:716-724.* DOI: 10.1002/hep.1840120416

[324] Fernandez J, Navasa M, Planas R, Montoliu S, Monfort D, Soriano G, Vila C, Pardo A, Quintero E, Vargas V, Such J, Gines P, Arroyo V. Primary prophylaxis of spontaneous bacterial peritonitis delays hepatorenal syndrome and improves survival in cirrhosis. *Gastroenterology* 2007;133(3):818-824. DOI: 10.1053/j.gastro.2007.06.065

[325] Salerno F, Gerbes A, Gines P, Wong F, Arroyo V. Diagnosis, prevention and treatment of the hepatorenal syndrome in cirrhosis. A consensus workshop of the international ascites club. *Gut* 2007;56:1310-1318.

[326] Angeli P, Merkel C. Pathogenesis and management of hepatorenal syndrome in patients with cirrhosis. *J Hepatol* 2008;48(Suppl.1):S93-S103. DOI: 10.1016/j.jhep.2008.01.010

[327] Navasa M, Feu F, Garciapagan JC, Jimenez W, Llach J, Rimola A, Bosch J, Rodes J. Hemodynamic and humoral changes after liver transplantation in patients with cirrhosis. *Hepatology* 1993;17:355-360. DOI: 10.1002/hep.1840170302

[328] Angeli P, Gines P. Hepatorenal syndrome, meld score and liver transplantation: An evolving issue with relevant implications for clinical practice. *J Hepatol* 2012;57(5):1135-1140. DOI: 10.1016/j.jhep.2012.06.024

[329] Bahirwani R, Campbell MS, Siropaides T, Markmann J, Olthoff K, Shaked A, Bloom RD, Reddy KR. Transplantation: impact of pretransplant renal insufficiency. *Liver Transpl* 2008;14(5):665-671. DOI: 10.1002/lt.21367

[330] Rossle M, Gerbes AL. TIPS for the treatment of refractory ascites, hepatorenal syndrome and hepatic hydrothorax: a critical update. *Gut* 2010;59(7):988-1000. DOI: 10.1136/gut.2009.193227

[331] Hadengue A, Gadano A, Moreau R, Giostra E, Durand F, Valla D, Erlinger S, Lebrec D. Beneficial effects of the 2-day administration of terlipressin in patients with cirrhosis and hepatorenal syndrome. *J Hepatol* 1998;29(4):565-570. DOI: 10.1016/S0168-8278(98)80151-7

[332] Ortega R, Gines P, Uriz J, Cardenas A, Calahorra B, De Las HD, Guevara M, Bataller R, Jimenez W, Arroyo V, Rodes J. Terlipressin therapy with and without albumin for patients with hepatorenal syndrome: results of a prospective, nonrandomized study. *Hepatology* 2002;36:941-948.

[333] Møller S, Hansen EF, Becker U, Brinch K, Henriksen JH, Bendtsen F. Central and systemic haemodynamic effects of terlipressin in portal hypertensive patients. *Liver* 2000;20(1):51-59. DOI: 10.1034/j.1600-0676.2000.020001051.x

[334] Mulkay JP, Louis H, Donckier V, Bourgeois N, Adler M, Deviere J, Le Moine O. Long-term terlipressin administration improves renal function in cirrhotic patients with type 1 hepatorenal syndrome: a pilot study. *Acta Gastroenterol Belg* 2001;64 (1):15-19.

[335] Moreau R, Durand F, Poynard T, Duhamel C, Cervoni JP, Ichai P, Abergel A, Halimi C, Pauwels M, Bronowicki JP, Giostra E, Fleurot C, Gurnot D, Nouel O, Renard P, Rivoal M, Blanc P, Coumaros D, Ducloux S, Levy S, Pariente A, Perarnau JM, Roche J,

Scribe-Outtas M, Valla D, Bernard B, Samuel D, Butel J, Hadengue A, Platek A, Lebrec D, Cadranel JF. Terlipressin in patients with cirrhosis and type 1 hepatorenal syndrome: A retrospective multicenter study. *Gastroenterology* 2002;122(4):923-930. DOI: 10.1053/gast.2002.32364

[336] Guevara M, Gines P, Fernandezesparrach G, Sort P, Salmeron JM, Jimenez W, Arroyo V, Rodes J. Reversibility of hepatorenal syndrome by prolonged administration of ornipressin and plasma volume expansion. *Hepatology* 1998;27:35-41. DOI: 10.1002/hep.510270107

[337] Duvoux C, Zanditenas D, Hezode C, Chauvat A, Monin JL, Roudot-Thoraval F, Mallat A, Dhumeaux D. Effects of noradrenalin and albumin in patients with type I hepatorenal syndrome: a pilot study. *Hepatology* 2002;36(2):374-380. DOI: 10.1053/jhep.2002.34343

[338] Pomier-Layrargues G, Paquin SC, Hassoun Z, Lafortune M, Tran A. Octreotide in hepatorenal syndrome: A randomized, double-blind, placebo-controlled, crossover study. *Hepatology* 2003;38(1):238-243. DOI: 10.1053/jhep.2003.50276

[339] Angeli P, Volpin R, Gerunda G, Craighero R, Roner P, Merenda R, Amodio P, Sticca A, Caregaro L, Maffei-Faccioli A, Gatta A. Reversal of type 1 hepatorenal syndrome with the administration of midodrine and octreotide. *Hepatology* 1999;29(6):1690-1697. DOI: 10.1002/hep.510290629

[340] Alessandria C, Debernardi-Venon W, Carello M, Ceretto S, Rizzetto M, Marzano A. Midodrine in the prevention of hepatorenal syndrome type 2 recurrence: A case-control study. *Dig Liver Dis* 200941(4).298-302.

[341] Wong F, Pantea L, Sniderman K. Midodrine, octreotide, albumin, and TIPS in selected patients with cirrhosis and type 1 hepatorenal syndrome. *Hepatology* 2004;40(1):55-64. DOI: 10.1002/hep.20262

[342] Alqahtani SA, Fouad TR, Lee SS. Cirrhotic cardiomyopathy. *Semin Liver Dis* 2008;28(1):59-69. DOI: 10.1055/s-2008-1040321

[343] Luca A, Feu F, Garcia-Pagan JC, Jiménez W, Arroyo V, Bosch J, Rodes J. Favorable effects of total paracentesis on splanchnic hemodynamics in cirrhotic patients with tense ascites. *Hepatology* 1994;20:30-33.

[344] Ruiz del Arbol L, Monescillo A, Jimenez W, Garcia-Plaza A, Arroyo V, Rodes J. Paracentesis-induced circulatory dysfunction: Mechanism and effect on hepatic hemodynamics in cirrhosis. *Gastroenterology* 1997;113:579-586. DOI: 10.1053/gast.1997.v113.pm9247479

[345] Møller S, Bendtsen F, Henriksen JH. Determinants of the renin-angiotensin-aldosterone system in cirrhosis with special emphasis on the central blood volume. *Scand J Gastroenterol* 2006;41(4):451-458. DOI: 10.1080/00365520500292962

[346] Bernardi M, Domenicali M. The renin-angiotensin-aldosterone system in cirrhosis. In: Ginés P, Arroyo V, Rodes J, Schrier RW, eds. *Ascites and renal dysfunction in liver disease.* 2nd ed. Malden: Blackwell Pullishing Ltd.; 2005. p. 43-54. DOI: 10.1002/9780470987476.ch4

[347] Gentilini P, Romanelli RG, La Villa G, Maggiore Q, Pesciullesi E, Cappelli G, Raggi VC, Foschi M, Marra F, Pinzani M, Buzzelli G, Laffi G. Effects of low-dose captopril on renal hemodynamics and function in patients with cirrhosis of the liver. *Gastroenterology* 1993;104:588-594.

[348] Schneider AW, Kalk JF, Klein CP. Effect of losartan, an angiotensin II receptor antagonist, on portal pressure in cirrhosis. *Hepatology* 1999;29(2):334-339. DOI: 10.1002/hep.510290203

[349] Gonzalez-Abraldes J, Albillos A, Banares R, del Arbol LR, Moitinho E, Rodriguez C, Gonzalez M, Escorsell A, Garcia-Pagan JC, Bosch J. Randomized comparison of long-term losartan versus propranolol in lowering portal pressure in cirrhosis. *Gastroenterology* 2001;121(2):382-388. DOI: 10.1053/gast.2001.26288

[350] Wong F, Liu P, Blendis L. The mechanism of improved sodium homeostasis of low-dose losartan in preascitic cirrhosis. *Hepatology* 2002;35(6):1449-1458. DOI: 10.1053/jhep.2002.33637

[351] Guyader D, Patat A, Ellis-Grosse EJ, Orczyk GP. Pharmacodynamic effects of a non-peptide antidiuretic hormone V2 antagonist in cirrhotic patients with ascites. *Hepatology* 2002; 36(5):1197-1205. DOI: 10.1053/jhep.2002.36375

[352] Fagundes C, Gines P. Hepatorenal Syndrome: A severe, but treatable, cause of kidney failure in cirrhosis. *Am J Kidney Dis* 2012;59(6):874-885. DOI: 10.1053/j.ajkd.2011.12.032

[353] Claria J, Jimenez W, Arroyo V, La Villa G, Lopez C, Asbert M, Castro A, Gaya J, Rivera F, Rodes J. Effect of V-1-vasopressin receptor blockade on arterial pressure in conscious rats with cirrhosis and ascites. *Gastroenterology* 1991;100:494-501.

[354] Jonassen TE, Christensen S, Marcussen N, Petersen JS. Intrarenal octreotide treatment prevents sodium retention in liver cirrhotic rats: evidence for direct effects within the thick ascending limb of Henle's loop. *Am J Physiol Renal Physiol* 2006;291(3):F537-F545. DOI: 10.1152/ajprenal.00226.2005

[355] Wong F, Blei AT, Blendis LM, Thuluvath PJ. A vasopressin receptor antagonist (VPA-985) improves serum sodium concentration in patients with hyponatremia: A multicenter, randomized, placebo-controlled trial. *Hepatology* 2003;37(1):182-191. DOI: 10.1053/jhep.2003.50021

[356] Tanaka M, Wanless IR. Pathology of the liver in Budd-Chiari syndrome: Portal vein thrombosis and the histogenesis of veno-centric cirrhosis, veno- portal cirrhosis, and large regenerative nodules. *Hepatology* 1998;27:488-496. DOI: 10.1002/hep.510270224

[357] Nicolau C, Bru C, carreras E, Bosch J, Bianchi L, Gilabert R, Vilana R. Sonographic diagnosis and hemodynamic correlation in veno-occlusive disease of the liver. *J Ultrasound Med* 1993;12:437-440.

[358] Plessier A, Valla DC. Budd-Chiari syndrome. *Semin Liver Dis* 2008;28(3):259-269. DOI: 10.1055/s-0028-1085094

[359] Darwish MS, Plessier A, Hernandez-Guerra M, Fabris F, Eapen CE, Bahr MJ, Trebicka J, Morard I, Lasser L, Heller J, Hadengue A, Langlet P, Miranda H, Primignani M, Elias E, Leebeek FW, Rosendaal FR, Garcia-Pagan JC, Valla DC, Janssen HL. Etiology, management, and outcome of the Budd-Chiari syndrome. *Ann Intern Med* 2009;151 (3):167-175. DOI: 10.7326/0003-4819-151-3-200908040-00004

[360] Ochs A, Rossle M, haag K, Hauenstein KH, Deibert P, Siegerstetter V, Huonker M, Langer M, Blum HE. The transjugular intrahepatic portosystemic stent-shunt procedure for refractory ascites. *N Engl J Med* 1995;332(18):1192-1197. DOI: 10.1056/NEJM199505043321803

[361] Møller S, Dumcke CW, Krag A. The heart and the liver. *Expert Rev Gastroenterol Hepatol* 2009; 3(1):51-64. DOI: 10.1586/17474124.3.1.51

Author Biographies

JENS H. HENRIKSEN

Jens H. Henriksen, M.D., D.M.Sc., R. is Chief Physician and Full Professor of Clinical Physiology at the Department of Clinical Physiology and Nuclear Medicine, Hvidovre Hospital, at the University of Copenhagen. He has served as Editor of *The Scandinavian Journal of Clinical and Laboratory Investigation*, Review Articles Editor of *Clinical Physiology*, and has been a member of several Editorial Boards. Dr. Henriksen is President of the Society for Theoretical and Applied Therapy and President of the Danish Society of Clinical Physiology and Nuclear Medicine. He has received several Awards (Kommunitetet, Tode Award, Klein Prize, and Niels A. Lassen Prize) and was recently knighted. His main fields of research are haemodynamic pathophysiology of liver disease (with special reference to circulatory dynamics, fluid homeostasis, tracer kinetics, and neuroendocrine regulation), kinetics models for protein and solute transport between different compartments of the body (including analysis of permeability and lymphatic transport), circulatory dynamics, fluid distribution, vascular tonus, and homoeostatic dysregulation. He has published a total of 450 scientific peer-reviewed papers, reviews, book chapters, and books.

SØREN MØLLER

Søren Møller received his M.D. and D.M.Sc. from the University of Copehnhagen, Denmark in 1987 and 1997, respectively, followed by becoming a Specialist in Clinical Physiology and Nuclear Medicine in 1998. Since 1999 he has been the Chief Physician at the Department of Clinical Physiology and Nuclear Medicine, Center of Functional and Diagnostic Imaging and Research, Hvidovre Hospital at the University of Copenhagen, Denmark. Since 2006, Dr. Møller has been an Associate Clinical Research Professor at the University of Copenhagen. He has published approximately 210 international and national publications including original papers, review articles, and book chapters on haemodynamic, humoral, and metabolic aspects of chronic liver disease. He has given more than 150 oral and poster presentations at international conferences and meetings. Dr. Møller was a member of EASL Scientific Committee from 1999–2002, a member of Board of the Danish Association for the Study of the Liver from 2000–2003, and served as President of the Danish Association for the Study of the Liver from 2003–2006. He was an UEMS and EANM representative of the Danish Society of Clinical Physiology and Nuclear Medicine from 2002–2005. From 2006–2011, Dr. Møller was a member of the Scientific Committee of the International Club of Ascites and served as President of the Danish Society for Clinical Physiology and Nuclear Medicine from 2006–2010. Since 2012 he has been a Professor of Clinical Physiology and Nuclear Medicine at the Faculty of Health Sciences of the University of Copenhagen, Denmark.